EVOLUTION IN MINUTES

DARREN NAISH

EVOLUTION IN MINUTES

DARREN NAISH

Quercus

CONTENTS

Introduction

Evolution by natural selection is the foundational concept in our study of the natural world. One of the most significant scientific discoveries ever made, it has revolutionized every aspect of our understanding both of the past, and of the diversity of living things today. It even allows us to make predictions about organisms that will evolve in the future. Indeed, evolution has phenomenal explanatory power: to quote biologist Theodosius Dobzhansky, 'nothing makes sense in biology except in the light of evolution.'

Evidence for evolution is everywhere, and studies show how the processes and events that together drive evolutionary change are continually present in every single community of living things. The result is that at least some knowledge of evolution is essential for anyone interested in the natural world. Nevertheless, the actual mechanism is complex and involves numerous different concepts, processes and events. New discoveries are constantly made. In fact, evolutionary science is often at the forefront of the news as advances in genetics, new fossil discoveries, and new data pertaining to the link between genetics and development are announced.

Some people, however, contend that the fact of perpetual change in nature challenges the words of certain sacred texts. For this reason, evolution is often controversial and even politicized. Religious leaders, politicians, celebrities and artists routinely comment on the topic, and it has been at centre stage in court cases and in arguments about education and public policy. The battle to teach children about evolution remains a pertinent one in some parts of the world.

In this book, we look at the key evidence for evolution, and the developing ideas of the 19th and 20th centuries. We investigate the principles, concepts and trends within evolution, and see how it has affected the history of life on geological timescales, and shaped the development of our own species. We end by considering the future of evolution: human-caused change, genetic modification and the possible role of cybernetics. The idea that living things change over time is basic, elegant in its simplicity, crucial, and fascinating, but there is still much to learn.

What is evolution?

The term 'evolution' relates to heritable change that occurs in living organisms over generations. That is, to change that occurs as one generation gives rise to the next. Because evolution take place on this gradual, generational scale, changes are often tiny and extremely slow to accrue or become obvious. However, this is not always the case. In organisms that grow and reproduce quickly, evolution can be rapid enough that changes are observed over the course of decades, or even years.

Evolution should not be confused with metamorphosis – the process by which living things change during the course of their lifetime. Nor, when we discuss biological evolution, are we discussing such things as the origin of life itself, or cosmological events, such as the creation of Earth or the Big Bang. The process of evolution is typically associated with the changes that affect natural populations of living things. However, organisms modified by people also evolve. In such cases, it can be said that humans control or shape their evolution (see page 26).

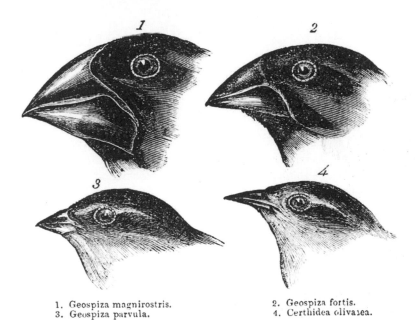

1. Geospiza magnirostris.
2. Geospiza fortis.
3. Geospiza parvula.
4. Certhidea olivasea.

Different bill shapes among Galapagos finches
offered the young Charles Darwin a vital clue to
the evolutionary mechanism.

The evidence for change

The idea that organisms change over the generations is supported by four facts. The first is that some organisms are similar to others, yet different enough that they have alternative lifestyles and do not interbreed. Zebras are similar to horses, for example. The second fact is that humans have modified many organisms – flowering plants, dogs and so on – by selecting individuals to breed with one another in order to emphasize specific features. If humans can cause such changes to occur over the generations, natural processes can too.

The third fact is that fossils show how living things have become smaller or larger over time, and how their parts changed shape or took on new functions. The fourth fact is that change over the generations has been observed in laboratory and natural settings. Scientists have observed generational changes in microscopic organisms and also in insects, fish, mice, lizards and birds. In some cases, the changes have been significant enough that we can probably talk about witnessing the origin of a new species.

The origin of species

If evolution refers to heritable changes that occur across generations, then it mostly concerns changes that happen on a small scale – in, say, members of a single population of a single species. This is the kind of evolution that scientists have most frequently observed. However, if such changes *are* happening, it follows that, should they continue for a sufficient period of time, a population will eventually become distinct enough from related populations that a new species will have emerged, or originated.

This issue formed the focus of Charles Darwin's pioneering book of 1859, titled *The Origin of Species by Means of Natural Selection, or the Preservation of Favoured Races in the Struggle for Life*. Darwin's model for the origination of species does not pertain to the origin of life as a whole, nor did Darwin understand how the information inherited by organisms was passed down across the generations. Substantial additional research was inspired by his proposal and continues today. Darwin's research and reasoning are also themselves a major area of modern investigation.

Survival of the fittest

The mantra 'survival of the fittest' is often mentioned as a key concept of evolutionary theory. Here, the term 'fittest' does not strictly refer to athleticism, strength or health. Instead, it refers to how good an organism is at surviving in its environment. The fittest organisms are those that are best at consistently finding food, water and shelter, the best at escaping predators, and the best able to produce surviving offspring. In other words, it describes the process of natural selection (see page 64). Over time, fitter organisms out-survive or out-compete less fit ones, the latter passing into extinction while the fitter ones produce successful descendants that inherit their high-fitness features.

The phrase 'survival of the fittest' is often said to have originated with Darwin. In fact it did not, but was instead first used by philosopher and biologist Herbert Spencer in an 1864 summary of Darwin's theory. Darwin did not object to this shorthand way of describing the process of natural selection and used it himself in some of his later writings.

Adaptation and mutation

Organisms change over time. Some of these changes make no difference to the success or failure of the organisms that inherit them; other changes do. Take a group of antelopes in which one individual has longer legs than its siblings. Those longer legs make the individual better at escaping from predators, hence giving it an advantage in the survival stakes. Its descendants inherit this feature and are better able to survive as a consequence. Those longer legs have thus become an 'adaptation'. Their evolution means that this particular lineage of antelopes has 'adapted' to its conditions, thanks to its development of this feature. But why do novel features appear in the first place?

Over the generations, the individual components of the genetic code within a species change, or mutate (see page 26). So far as we know, mutations may provide no evolutionary advantage at all. They may be disadvantageous, in which case they are weeded out via natural selection; or they may be advantageous, in which case they persist through the generations.

Chicken or egg?

If evolution involves one species evolving from another, at which point do we say that a new species has emerged? Can a chicken (for example) lay an egg containing the first member of a new species? Experts disagree over the speed at which evolution occurs. However, even those who endorse the fastest version of evolution imaginable think that it works on the scale of many generations. If we imagine watching one species evolve into another, we would see white turn to black via innumerable shades of grey.

Species are human-made constructs: we agree that a 'species' is a population that possesses traits that distinguish it from both ancestral and descendant species. As a result, there must be one particular 'shade of grey' – one particular generation in an evolutionary sequence – that has the trait (or traits) that make it a member of species *x*, instead of species *y*. The result is that, yes, there must have been events in which a given set of parents gave rise to offspring that were the first of a new species.

Am I a monkey?

It's obvious that humans are primates. Within living primates, it's also clear that humans are closely related to the African apes, in particular, the chimpanzee and bonobo. In fact, humans are part of the African ape group, in which case it is accurate to describe humans as apes in the same way that chimps and bonobos are.

Apes belong to a group of primates (termed 'anthropoids') that also includes New World monkeys, such as spider and howler monkeys, and Old World monkeys, such as baboons, macaques and colobus monkeys. No New World monkeys are ancestors of apes. Old World monkeys, however, share an ancestor with apes that would almost definitely be considered a monkey were it alive today. Furthermore, 'monkey' is not a formal zoological term: traditionally, it merely refers to any anthropoid that is not an ape. But if apes descend from a primate that we would term a 'monkey', it is correct to describe apes as a particular group of large-bodied, tailless monkeys. According to this view, humans – like all apes – are indeed giant, modified monkeys.

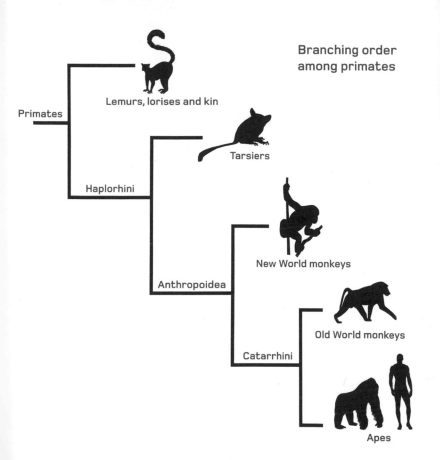

Branching order among primates

Primates

Lemurs, lorises and kin

Haplorhini

Tarsiers

Anthropoidea

New World monkeys

Catarrhini

Old World monkeys

Apes

Sex and inheritance

People knew for hundreds of years prior to the work of Darwin that offspring inherited the traits of their parents. They took advantage of this knowledge by crossing domestic plants and animals with one another in order to accentuate certain traits, and reduce or lose others. Exactly how the transmission of information across the generations occurred was mysterious and not really resolved until the 20th century.

Why two parents are involved in the formation of offspring is a good question, especially when there are large numbers of living things (microorganisms and various plants and animals) that reproduce asexually – that is, without a partner. Sexual reproduction involves the combination of genes from two individuals. Offspring are not copies of either parent, but have their own genetic code. Why this is advantageous remains the topic of debate, but the most popular idea is that the increased variation seen in the offspring of sexually reproducing organisms allows for swifter adaptation and thus faster evolution.

Genes and DNA

Information passed across the generations does so via the molecule DNA and the individual segments – termed 'genes' – included within it. While DNA was not fully understood to be key to inheritance until the 1950s, genes were linked to inheritance some time earlier. During the 1920s, experiments performed by biologist Frederick Griffith showed that genes could be transferred. Prior to these studies there was no clear idea on how heritable information was transmitted.

DNA (deoxyribonucleic acid) is an elongate molecule consisting of two strands formed of the sugar deoxyribose and phosphate radicals. Each unit of deoxyribose is connected to a biological compound termed a 'nucleobase', and four nucleobases occur: guanine, cytosine, adenine and thymine (G, C, A and T). A sugar, phosphate and nucleobase together form a 'nucleotide'. The two strands of DNA are linked via hydrogen bonds that connect the nucleobases, the only bonds possible being those between guanine and cytosine, and between adenine and thymine.

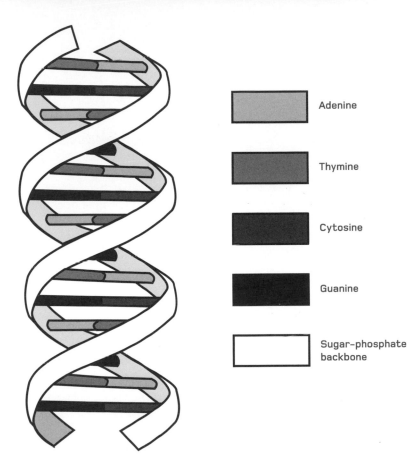

Adenine

Thymine

Cytosine

Guanine

Sugar-phosphate
backbone

Descent with modification

Organisms produce descendants, and these descendants
– their immediate offspring and subsequent generations
– inherit the genetic code of their ancestors. If this is all that
happens, evolution would not exist. But as genetic codes are
copied across the generations, changes occur. These changes
(termed 'mutations') result in variation, and the accruing of
sufficient variation over time results in generations that have
features quite different from those of their ancestors.

Combined with the phenomenon of natural selection (see page
64), the concept of descent with modification is central to
the idea of evolution. The fact that change over generations
is such a key concept is important: evolution does not, by
definition, involve the changes that happen to individual
organisms across their lifetime.

Also significant is the fact that any changes are inherited, and
thus passed down from one generation to the next.

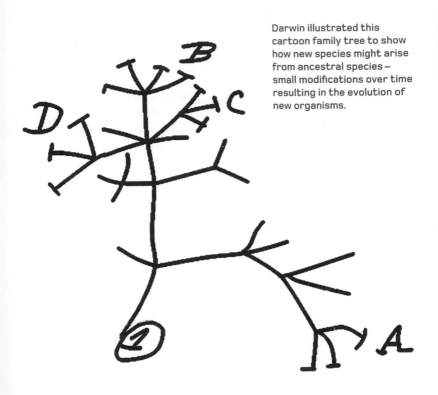

Darwin illustrated this cartoon family tree to show how new species might arise from ancestral species – small modifications over time resulting in the evolution of new organisms.

Long before Darwin

The discovery of evolution is famously associated with Charles Darwin, but he was not the first to consider the idea. During the 18th and 19th centuries, several biologists thought it likely that living things had changed over time. These people were typically somewhat radical in social or political terms and were not afraid to challenge the prevailing view that all living things were created according to a grand, divine plan. In England, Erasmus Darwin – grandfather of Charles – alluded to evolution in his writings of the late 1700s. In France, Jean-Baptiste de Lamarck endorsed a view of evolution whereby the habits of living things caused them to change (see page 46).

Lamarck's colleague Étienne Geoffroy Saint-Hilaire proposed that environmental changes caused living things to become modified via sudden events. Other scientists across continental Europe adopted similar views, and models involving transformation – often involving the modification of an ancestral plan – were familiar well before Darwin published his *Origin of Species*.

Erasmus Darwin

The discovery of fossils

Fossils – the remains of once-living things preserved within Earth's rock layers – are commonplace in many parts of the world, most typically on eroding coastlines and in deserts where exposed rock is worn away by wind and sand. The ancient Greeks and others knew of fossils and realized that they were similar to the skeletons of modern animals, though they seemingly interpreted them as the remains of mythical beasts and heroes.

By the 19th century it was well understood that fossils were the petrified remains of living things from long ago, though the actual ages involved remained the subject of debate and guesswork. Dinosaurs, great sea reptiles, giant fossil mammals and fossil relatives of humans were all known by the mid-1800s. These fossils were clearly consistent with emerging ideas about evolution, though the numerous gaps in the fossil record meant that many transitions between groups remained unknown. Darwin and others of his time hoped that these gaps would eventually be filled. And, eventually, many were.

18th-century scientist Georges Cuvier compared fossil forms to those of living creatures.

Snakes with legs and walking whales

Darwin predicted that, eventually, a great number of 'transitional fossils' would be discovered: that is, fossils representing species intermediate in anatomy between major groups. One of the first sets to be documented concerns the origin of mammals from reptile-like ancestors.

Numerous species – mostly dating to the Permian – chart transformations that occurred in ear, tooth, jaw and brain anatomy as mammals emerged. Since the 1980s several fossil sea cows and whales with well developed limbs and an ability to walk on land have been discovered, as have species that link the earliest tetrapods (see page 278) with their fleshy-finned fish ancestors. We also know of fossil snakes (mostly from the Cretaceous) that have small limbs and digits, and of numerous ancient sea creatures that link the earliest vertebrates with their vaguely worm-like relatives. Although far from complete, the fossil record is now substantial enough that numerous transitional events are well understood.

Ambulocetus was an ancient member of the whale family, yet it had four well-developed limbs and could clearly walk on land.

Discovering geological time

Key evidence for evolution comes from living organisms; even if fossils and rocks did not exist we would still conclude that evolution occurs. However, there is no denying that the discovery of geological time – that Earth existed for vast periods prior to the modern age – helped people come to terms with the idea that living things have had plenty of time to change.

British scientists began to consider geological time during the 18th century. In Scotland, geologist James Hutton pointed to evidence that rock layers had been laid down over time, that some layers had been destroyed and new ones deposited, and that there was evidence for processes just like those of modern times operating in the past. His famous quote from 1788 is 'we find no vestige of a beginning, no prospect of an end' in this geological record. The world had not been created in a brief moment. Other geologists confirmed and backed up Hutton's observations. The next challenge was to pin actual ages – in decades, centuries, millennia or more – on to specific rock layers.

Dating the past

Evidence from the geological record led scientists to realize that rock layers were laid down over vast spans of time. But how can actual ages be determined for these rock layers, and for the fossils they contain?

Several structures form on a year-by-year basis and hence record actual time. Examples include tree rings, layers in sediment termed 'varves', and layers in ice. However, these structures 'only' record time as far back as 800,000 years or so. During the 1940s scientists demonstrated that radioactive elements such as carbon-14 (a particular form, or isotope, of carbon) decay at a predictable rate over time, and that their decay is constant following the point at which the organic object being studied stopped exchanging chemicals with the atmosphere. This technique is termed 'radiometric dating'. Carbon-14 only provides data back to about 60,000 years, but other radioactive isotopes decay far more slowly and can be used for dating much older samples.

Radiometric dating measures the ratio of radioactive 'parent' isotopes relative to the daughters they form over time. Half the atoms of a particular isotope will decay into daughters with each 'half life', so the balance between parents and daughters will shift.

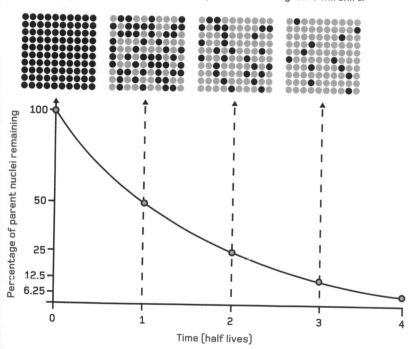

The fossil record

Most people are aware of the existence of fossils and that they provide evidence for evolutionary change. But few non-scientists have much grasp as to how extensive the fossil record is, or of the quality of evidence it represents. In fact, the record of ancient life, preserved in rocks, is outstandingly good.

The fossil record of younger rocks is better than that of older rocks, which have endured more weathering. Furthermore, some organisms are frequently not preserved, and there are environments in which given groups were rarely incorporated into the rock record. Despite these facts, the sheer number of fossils is staggering: millions of ancient seagoing planktonic creatures, trilobites, ammonites and other shelled invertebrates have been discovered and collected. Fossils belonging to these groups are so abundant, and so continuously preserved, layer upon layer, that they reveal obvious evidence for evolution, and show how species change via tiny increments as we move upwards in the rock sequence.

Linnaeus and the classification of life

Carl von Linné – often known by the Latinized Carolus Linnaeus – was an exceptional Swedish biologist especially interested in botany. During the 1730s, Linnaeus developed a catalogue for the identification of plants. By the 1750s he had developed a standardized system for the naming of organisms: they were given a two-part name (a binomial) consisting of an initial genus name (intended to include any number of similar species) and a specific or trivial name. By 1758 and the time of its tenth edition, Linnaeus' catalogue – the *Systema Naturae* – included over 12,000 species. His binomial system became mainstream, as did the groupings he favoured for organisms.

Linnaeus did not endorse an evolutionary perspective, but arranged organisms according to what he regarded as God's plan. However, his demonstration that organisms could be arranged into sets, each of which were then arranged close to other sets, helped establish the fact that all life was connected in a pattern indicative of evolution.

Cuvier and the discovery of extinction

We know today that extinction events occurred in the past and that species become extinct all the time, but such ideas were not accepted until the late 1700s. Prior to this time, statements in the Bible were thought to demand that living things were immutable: brought into existence at the dawn of life, they would persist for eternity. This concept of immutable species was inconsistent with the evidence for evolution, since the persistence of species forever disallows the appearance of new ones.

Fossils of large vertebrates led scientists to consider the reality of extinction. The leading anatomist and zoologist of the late 1700s and early 1800s was Georges Cuvier. Highly skilled in the interpretation of fragmentary fossils, he examined elephant fossils found near Paris in the 1790s, and argued that they simply must represent species no longer present on Earth. Other fossils – the first pterosaur and a giant swimming lizard called a mosasaur – were identified by Cuvier in the early 1800s. These discoveries established the fact of extinction.

The giant swimming lizard *Mosasaurus* was one of the key discoveries used by Cuvier to document the existence of extinction.

Anatomy and homology

If living things have changed over time, it follows that their anatomical structures represent modified versions of older structures. But sometimes these modifications appear radically different to the ancestral structures: a limb originally used in walking becomes a wing or a fin, for example; or a tongue used in manipulating food in the mouth becomes an extensible object that grabs food some distance away. We term such structures 'homologues'; they are 'homologous'.

Some examples of homology are obvious. The wings of birds, bats and the extinct pterosaurs, for example, are homologous: all are composed of the same set of tissues and emerge from the same part of the body of the developing embryo. Homology is important because it allows us to determine that organisms that look distinct and function in very different ways still share an evolutionary origin. Fossil, genetic and embryological discoveries can also show that organisms are *not* as closely related as might otherwise be thought.

Wing Homology

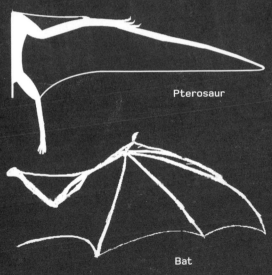

Bird

Pterosaur

Bat

These three creatures represent different groups — bird, reptile, mammal — yet their wing structures are homologous.

Lamarck and Lamarckian evolution

The theory of evolution by way of natural selection describes how the outcome of genetic variation is shaped by selection. Several other models of evolution have been proposed though, among the most persistent of which is Lamarckian evolution. Jean-Baptiste Lamarck argued that change in organisms was driven by deliberate choice: that they modified their habits by will or effort, and that their biology and anatomy changed in step with these decisions. Here's a popular example of how this might work: if, every generation, giraffes craned their necks upwards in a continual effort to reach ever higher, so their necks – Lamarck proposed – would gradually increase in length.

Today we know that this concept of evolution is erroneous. Giraffes did not grow long necks because they stretched a lot, but because natural selection favoured the survival of individuals with longer necks. that arose through genetic variation. Lamarckian evolution was surprisingly popular until the middle of the 20th century, especially in France.

Mendelian genetics

Gregor Mendel was an Augustinian friar based in what is now the Czech Republic. By breeding and growing plants of several sorts under experimental conditions – most famously peas – he was able to show that characteristics were passed from one generation to the next via inheritance.

General thinking at the time was that parental features simply became mixed or blended as they were passed to descendant generations. Mendel showed, via simple experiments involving plants of differing height or colour, that some traits were recessive (appearing several generations down the line), while others were dominant (appearing immediately in the next generation). Published in 1866, his work was key in showing that invisible factors – now known to be 'genes' – carried information across the generations. Despite the significance of Mendelian genetics in establishing the mechanism and nature of inheritance, Mendel's work was ignored for decades and not really appreciated until the beginning of the 20th century.

Alfred Russel Wallace

Alfred Russel Wallace was a schoolmaster and surveyor who set out, in 1848, to collect natural history specimens from the Amazon. In 1854 he travelled to the Malay archipelago. Both trips led him to realize that living things changed over time, guided by natural selection. In other words, he discovered the same model of evolution as Darwin did, and his idea was presented in conjunction with Darwin's to the Linnaean Society of London in 1858. Care was taken to ensure that Darwin received credit for being first to devise this proposal. They remained on good terms despite the conflict that might have arisen and Wallace's book of 1889 is even titled *Darwinism*.

Wallace is best known for his work on animal distribution. During the 1870s and 1880s, he established that specific groups were only found in certain areas, and that extinction had removed animals from areas where they formerly occurred. This work was fundamental to biogeography, the study of how and why living things are distributed the way they are (see page 52).

The science of biogeography

Few living things occur worldwide. Instead, they are restricted to parts of Earth's surface, and there are places – corresponding to changes in climate or elevation or to geographical obstacles – at which their distribution ends. Today, we know that organisms have ranges that correspond to evolutionary and geological history. However, this important discovery only came about once people had begun to systematically survey organisms and their distribution.

The field of biogeography was developed during the late 1800s, largely from the work of Alfred Russel Wallace. In South East Asia, Wallace discovered a boundary where Australasian animals were located to the east and Asian animals to the west. The boundary – today known as Wallace's Line – reveals the distinct histories of the two assemblages of animals and shows that they have not yet had time or opportunity to cross the boundary. The distribution of some organisms was enigmatic in Wallace's time and it would take the discovery of continental drift to resolve such issues.

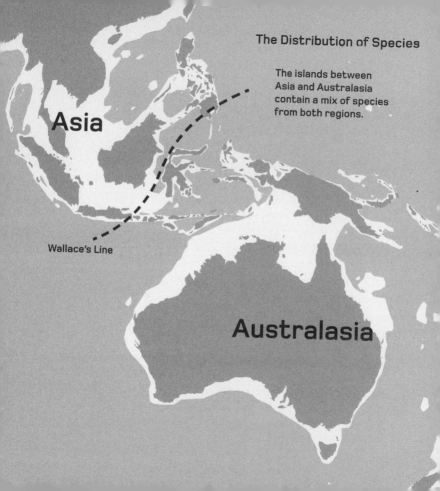

The Distribution of Species

The islands between Asia and Australasia contain a mix of species from both regions.

Asia

Wallace's Line

Australasia

Charles Darwin

English naturalist Charles Darwin is associated with evolutionary theory more than any other person. From the 1830s onwards, he collected information on animals, plants and fossils; by at least 1837, he believed that living things must have changed over time. In his 1859 book *The Origin of Species*, Darwin presented copious data from domestic animals and plants, from the distribution, biology and behaviour of living things and from the geological record, in order to discuss his theory of evolution by means of natural selection. It remains one of the most famous and influential books ever written.

Darwin gained crucial insight into the natural world during his voyage on HMS *Beagle* (see page 56). He also studied living things at home in England, including carnivorous plants, earthworms and even his own children. Most of his writing was done at Down House, his family home at Downe, Greater London. Darwin published numerous studies after *The Origin of Species*, focusing variously on orchids, domestication and the behaviour of worms.

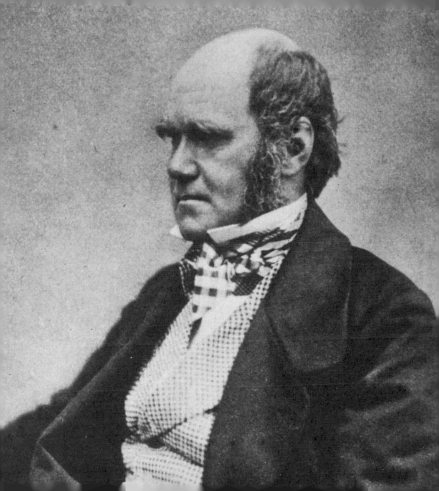

The voyage of the *Beagle*

In 1831, Charles Darwin was invited to serve as naturalist for the research vessel HMS *Beagle*. This voyage aimed to survey southern South America and would involve stops at Cape Verde, Brazil, the Falkland Islands, Argentina, Chile, the Galapagos Islands, Tahiti, New Zealand, Australia and South Africa. Darwin was ideal given his training as a naturalist with geological knowledge. He was just 22 years old at the time.

During the visit to the Galápagos Islands off the coast of Ecuador, Darwin encountered giant tortoises, seagoing iguanas and numerous other unique island-dwellers. These animals were close cousins of species on the South American mainland, yet island life had somehow made them different. On returning home, Darwin compiled his observations for a book – *The Voyage of the Beagle* (1839). This, combined with his writings on rocks and discoveries of fossils, made him well known in scientific circles. The data he collected, and things he saw, proved pivotal to the development of his ideas about evolution.

Domestication: human-made evolution

Throughout human history, humans have domesticated numerous plants and animals. Domestication involves the maintaining of a group of living things separately from 'the wild', and modifying its biology such that it becomes better suited for human use. Those uses may be as a food source, a labour-saving device, a companion or an ornament. In recent decades, new breeds of lizard and snake have been created for the pet trade and new breeds and strains of domestic dog, chicken and plant have been bred into existence.

Some of the changes arising from domestication have been profound, resulting in animals with new features not present in wild relatives. The fact that people have been able to modify the looks and behaviour of living things, merely by controlling which organisms get to breed together, shows how quickly major change can occur. Domestication is essentially 'human-made evolution', operating as an accelerated version of the sorts of changes that can occur via natural selection.

Discovering adaptation

Evolution does not work towards a goal or end point. Rather, natural selection acts as a 'shaper' of organisms, such that living things become better suited over time for a particular way of life. Eventually they become so well suited to that way of life that they cannot survive in any other way. We term this key process 'adaptation', and describe living things as being 'adapted' for a particular environment, habitat or way of life. Prior to Darwin's writings, it was often noted how organisms were superbly adapted for their environments. Darwin's key insight was to show that those adaptations must be the result of change such that the organism could, via natural selection, *remain adapted* to its changing environment.

As examples, Darwin pointed to wolf populations that differed in leg length and body shape according to the swiftness of the local prey available to them, and to the way in which the flowering parts of plants have changed according to the behaviour of the insects that pollinate them.

Darwin argued that powerfully built wolves would have an advantage in tackling heavy prey such as bison, while faster and slimmer wolves might be better able to capture smaller animals like deer.

Plants and the struggle for existence

Plants have been crucial in shaping ideas about evolution: generational changes in plants have provided information on hybridization and heredity, and plant adaptation and behaviour are integral to evolutionary theory. Plants are often more 'plastic' than animals, exhibiting abilities to change form according to conditions, and thus help show how evolution can and does occur.

Plants are not static or passive – it is common for species to compete for light and space. Many plants grow leaves that are larger and more horizontally aligned, solely to give them an advantage in out-competing neighbours. Of many hundreds of seeds and seedlings, only a handful may avoid the actions of plant-eating animals to reach maturity. Spines, toxic chemicals and camouflage have evolved to minimize risk. Darwin wrote about the factors that influence the survival and perpetuation of plants and described how they endured a 'struggle for existence'. This was a foundational concept for the development of his theory of natural selection.

The theory of natural selection

Darwin realized that living things were 'selected' by natural processes. The principle is simple: organisms are variable and numerous offspring perish before breeding. Those possessing heritable features that aid their survival and ability to reproduce are more likely to persist and produce successful offspring themselves. Specifically, Darwin described the process involved as that in which the useful variations of a feature are retained. He was inspired by his observations of domestic animals and the breeds created artificially by humans, but also drew on numerous examples he had witnessed in nature.

The processes of natural selection – predation, climate, food supply and so on – do not prevent unfit individuals from passing on their genes. Over long spans of time, however, unfit lineages are less likely to persist and tend to be out-competed by fitter, better-adapted lineages that are able to breed faster, to obtain, reach or use food sources more effectively and to avoid predators, harsh weather and so on.

The varied shell shapes of giant tortoises on different Galapagos Islands are evolutionary adaptations for different living conditions and food resources.

Sexual selection

One of the paradoxes of evolution is that sexual selection is in operation at the same time as natural selection, but often operates in contradiction to it. Sexual selection describes the process in which those organisms best able to pass on their genes out-compete those less able to do so. It explains why evolution favours the development of such things as display structures and risky, attention-grabbing behaviours. Yet, such sexually selected features make it harder for organisms to hide from predators, avoid parasites and find refuge from the weather – in other words, to escape the rigours of natural selection.

A classic example of sexual selection is the male peacock's extravagant tail feathers (which are not really part of the tail, but instead upper tail coverts). The demands of natural selection are such that the bird is slowed down and made highly visible thanks to these super-developed feathers. However, they increase its success in mating and thus have persisted owing to the pressure of sexual selection.

Niches and niche differentiation

Natural selection has shaped all organisms to become adapted to a specific way of life in a specific habitat – adapted to a 'niche'. The concept of the niche is thus key to adaptation and to evolution as a whole. Evolution has seen organisms adapt to fill most niches available, though some do remain unexploited.

The greatest number of niches occur in the richest, most fertile, most complex environments. This explains why a greater number of species, and more adaptation and variation, is present in such places. Because adapting to the same niche will result in direct competition, species are generally adapted to distinct niches, though the same niche may be occupied by different species if that niche provides sufficient opportunity. The process whereby closely related species have adapted to distinct niches is termed 'niche differentiation'. A caveat is that niche differentiation is not exclusively relevant to the diversification of species, since there are some organisms where members of a single species can occupy several distinct niches as it changes across its life cycle.

The extinct huia was a remarkable bird in which long-billed females and stout-billed males occupied different niches.

Ernst Haeckel's embryos

The embryos of many living things look substantially more alike than do the adults. The idea that this provides evidence for descent from common ancestors is famously associated with Ernst Haeckel, who drew attention to this fact in the 1860s.

Although Haeckel's name is almost synonymous with his work on embryos, he was highly accomplished, published numerous key ideas and discoveries, and did much to promote Darwin's writings. It turns out that Haeckel cheated somewhat in his work on embryos and made them look more alike than they actually were. None of this eliminates the main point that vertebrate embryos do look alike at key stages of development.

Haeckel thought that this was evidence that animals passed through the ancestral stages of their own evolution during their embryonic development, an idea termed 'recapitulation' and often paraphrased as 'ontogeny recapitulates phylogeny'. It is not accurate, however, and has long been discredited.

Fish

Reptile

Similarities present between vertebrate embryos point to shared ancestry and similar patterns of development.

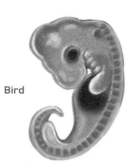

Bird

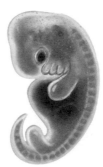

Human

Miescher and the 1860s discovery of DNA

While DNA was 'deciphered' by James Watson and Francis Crick in the early 1950s (see page 202), it was actually discovered some considerable time earlier.

During the late 1860s, Swiss physician Friedrich Miescher sought to determine the molecular structure of cells collected from human patients. The concept of cells themselves was a new one and any work determining their structure and function was exciting, cutting-edge science. Miescher was able to extract a newly discovered molecule from the cell nucleus – he termed it 'nuclein' – and to determine its composition. Later experiments on salmon cells confirmed its presence, which Miescher announced in a publication of 1871. He had discovered DNA. The exact significance of this discovery was not fully appreciated: Miescher did not link nuclein to heredity and instead thought that proteins were key to that process. Indeed the idea that proteins were key to the transmission of hereditary information remained the mainstream hypothesis until the 1940s.

Hopeful monsters

Of the many anatomical transformations that have occurred during the history of life, some appear to have been so weird or major that they involved strange 'halfway-house' creatures. In 1940, Germany geneticist Richard Goldschmidt (1878–1958) proposed that evolution might proceed via large saltations (abrupt variations) involving creatures that he termed 'hopeful monsters'. This name implies that such creatures must have been monstrous oddities, 'hopeful' in the sense that they might have survived and begat descendants, but – equally well – might not.

For a while it was thought that Goldschmidt's idea was flawed. Within recent decades, however, animal populations have been discovered where large 'jumps' of this sort have been discovered. In addition, fossil animals with unusual transitional features once thought impossible have been discovered. One example concerns fish that have a nostril on the edge of the upper jaw – a peculiar position unlikely to have improved the fish's ability to feed. The nostril later moved on to the palate.

A sudden switch in snake evolution

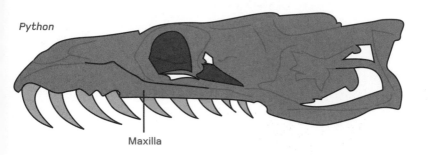

Python

Maxilla

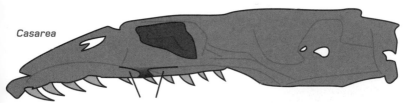

Casarea

Bipartite maxilla

Casarea belongs to a group of snakes where the maxilla – ordinarily a single bone – is split in two and has a mobile joint in the middle. It is almost impossible to think of a prototype version that would have given the animal an advantage. The change must, therefore, have occurred via a sudden saltation.

A mechanism for evolution

Darwin's key insight was to propose that natural selection was the primary mechanism of evolution. In developing this idea, he combined his observations of selective breeding and adaptation in organisms with the argument that increases in population would outstrip food supplies if all offspring survived. That argument about population increase had been put forward by the Reverend Thomas Malthus in an essay published in 1798. It was integral to Darwin's notion of survival of the fittest, since it was obvious that more organisms came into being than could ever survive.

Darwin defined natural selection as the 'principle by which each slight variation (of a given trait), if useful, is preserved'. He thus proposed a very different view from that promoted by Lamarck in which it was thought that the lifestyle of individuals was the key mechanism behind change (see page 46). The concept of natural selection proved, perhaps, the most controversial aspect of *The Origin of Species* after its publication, and Darwin himself wished that he had termed it 'natural preservation'.

Thomas Malthus' theories on the growth of populations over time were key to Darwin's ideas that only some members of a population could ever survive.

The Darwinian debate

Darwin's argument – put forth in *The Origin of Species* in 1859 – was not accepted by all of his colleagues and peers. Negative reviews appeared in several publications, and many scholars and members of the clergy expressed dismay that change might occur without the guiding hand of a creator; leading scientists of the time – Richard Owen and Thomas Huxley among them – argued over the evidence. In ways, the debate was less vociferous than it might have been, since an earlier volume – the anonymous *Vestiges of the Natural History of Creation* (1844) – had already proposed that natural processes had shaped the world more than religious leaders thought likely.

For all the negative things said about *Origin*, praise was evident too, and Darwin's ideas were widely known and regarded by many as likely to be correct by 1880. However, the possibility that evolution might occur via Lamarckian mechanisms became increasingly popular during the late 1800s. The idea that humans had emerged from 'lower animals' also remained contested.

[December 8, 1881.

MAN · IS · BVT · A · WORM

Thomas Huxley: Darwin's bulldog

Charles Darwin is famously associated with the theory of evolution, but so is Thomas Henry Huxley (1825–95), a British comparative anatomist who wrote and lectured extensively on evolution and biology. Huxley is well known today for his technical articles on dinosaurs and was one of the first scientists to point to the detailed similarities present between newly discovered fossil dinosaurs and birds. Prior to this, he was assistant surgeon aboard the HMS *Rattlesnake* and voyaged to Australia and New Guinea while in his 20s. At various times during his career, he studied invertebrates, the shared features of apes and humans, and vertebrate fossils.

Huxley did not immediately accept many of the ideas put forward by Darwin, but did nevertheless promote and support his work and the concept of evolution. His strong advocacy of evolution led to him being termed 'Darwin's bulldog' and he used his excellent skills in oration and argument to demolish opponents, most famously Bishop Samuel Wilberforce in a debate of 1860.

Bishop Wilberforce vs Thomas Huxley

Darwin's theory of evolution by way of natural selection was deemed controversial, one reason being that it removed the need for a god. Partly as a consequence of this mood, a debate took place at the Oxford University Museum in 1860. The most vocal in the debate were Thomas Huxley and Samuel Wilberforce, Lord Bishop of Oxford. Wilberforce had spoken out about evolution for years and had been coached by the great comparative anatomist Richard Owen, an opponent of Darwin.

While the debate did not concern Huxley and Wilberforce alone, the best remembered statements are those made by Huxley. In the most famous part of the conversation, Wilberforce asked Huxley whether it was on his mother's side or father's side that his ape ancestors were to be sought. The exact response is not recorded, but it was along the lines that Huxley would rather be descended from an ape than a man who misused his talents to suppress debate. This response was widely reported in the press and increased Huxley's fame and standing in educated circles.

Inner urges
and other ideas

At several times in history, people have thought that living
things were driven by some kind of internal urge to
improve themselves, to aim for a goal of some kind. Those
supporting this view pointed to evidence that living things had
evolved in straight-line fashion towards those given goals. This
idea is termed 'orthogenesis' and has several variants, including
aristogenesis, nomogenesis and heterogenesis. Popular in the
late 1800s and early 1900s, the idea persisted until the 1930s.

Orthogenesis is similar to Lamarckian evolution (where habitual behaviour is imagined as the main driver of change), and some scientists did see it that way. However, the concept of an almost mystical inner urge to achieve a goal and evolve directly towards it has also been key to orthogenetic ideas. Increasing acceptance of natural selection as the primary driver of change helped replace the idea, as did the fact that those lineages thought to demonstrate 'straight-line' or 'goal-driven' evolution did no such thing.

The Scopes Trial

Some parts of the world have seen social, political and religious movements emerge whereby people protest the teaching of evolution, their argument being that it contradicts or undermines the word of a chosen religious text. Parts of the predominantly Christian United States – almost uniquely among the developed nations – have often resisted efforts to have evolution taught in schools.

In 1925, Tennessee teacher John T. Scopes was accused of teaching evolution to his students, an act disallowed by state law. He was taken to court – the case being known as the Scopes Trial or Scopes Monkey Trial – and found guilty. The verdict was, however, overturned. The Scopes Trial is the most famous legal case involving evolutionary theory, but it was never really about Scopes specifically. Instead, it was deliberately orchestrated by the American Civil Liberties Union, so that lawyers could debate – and hopefully repeal – the banning of teaching about evolution. It ignited a huge amount of public debate.

Substitute teacher John Scopes could not recall whether he'd specifically mentioned evolution in his classes but cooperated with others to make sure that the trial went to court anyway.

Racism and evolution

Scientific ideas are devised and developed on the basis of information available at the time. Science does not, therefore, work to a given bias or agenda. In reality, the biases and agendas practised by various people through history have seen them distort scientific evidence to suit their own perspectives. By far the most notorious and damaging example concerns racism.

At times, given groups of people have regarded themselves as superior – intellectually, physically or technologically – to other groups, and claimed that scientific evidence supports this view. During the 1930s, the German Nazis argued that Nordic and Germanic people were the most superior examples of our species and that Jews and others were inferior and 'lower' in evolutionary terms. Other racist groups have also promoted the view that pale-skinned people are somehow 'more advanced' than dark-skinned people. None of these views match scientific evidence on human evolution, nor do they take account of what we actually know about human history, industry, art and culture.

Haeckel. Entwickelungsgeschichte

2. Gorilla.

1. Schimpanse.

3 Orang.

4. Neger.

Lith Anst v E.Giltsch.Jena.

Some early depictions of evolution tried to denigrate non-white humans by placing them closer to non-human apes on the evolutionary tree.

Creationism

Creationism is generally understood as the view that evolution has not occurred (or, at best, that major macroevolutionary events have not occurred; see page 138) and that living things were put into place by an all-powerful entity: typically, the Christian or Islamic god. Creationism usually involves a literal interpretation of a specific religious text, most typically the Bible or Qur'an, and comes from people who want religious teachings to be at the forefront of daily life and to take precedence over scientifically based views of the universe.

Given that virtually all religions involve the existence of a creator of some kind, it might be true to say that all interpretations of the universe that involve a creation event can be termed variants on creationism. Nevertheless, views involving literal interpretations of religious texts, a short duration between creation and the present, and an absence of macroevolution are typically thought to be key. There are, however, variants of creationism that allow for some evolution, or for a geologically old Earth.

Intelligent design

During the 1980s the term 'Intelligent Design' (ID) appeared in creationist literature. Its proponents have pushed for its teaching in schools, their claim being that ID is a viable alternative to the theory of evolution by way of natural selection, and that it is science-based and not religiously motivated.

ID predicates that complexity in nature can only have appeared due to the role of a designer – an intelligent entity working to a plan. The designer is clearly the Christian god, and ID proponents have stated such on occasion. The vertebrate eye is a favoured example of ID proponents, as is the whip-like tail structure (the flagellum) of some bacteria.

The realization that ID is a form of creationism has led to court battles in the United States and an effort to keep it out of classrooms. Its teaching in schools would be inconsistent with the United States Constitution, since this requires separation between church and state.

Intelligent Design theory argues that structures like the eye cannot have evolved over time. In fact, eyes are full of flaws and makeshift solutions – exactly what we would expect given the fact of evolution.

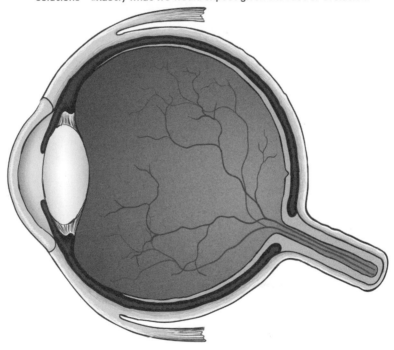

The march of progress

The theory of evolution is, in basic terms, relatively simple: offspring are variable, natural selection acts on those offspring, and the better-adapted ones produce more surviving offspring. While there are indications of 'improvement' and 'increasing fitness' within evolution, it is not a march towards some predetermined goal. While this fact is stated frequently, evolution is still imagined to be exactly this in popular culture.

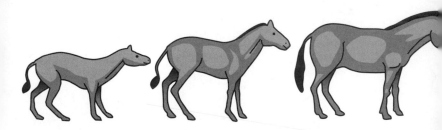

Authors and artists often depict evolution as a 'march of progress' in which organisms (most frequently primates or horses) have evolved directly from a primitive, lowly ancestor to an advanced, noble 'end product'. In some ways, these illustrations are not inaccurate: they usually do depict a trend or tendency within a lineage, but for reasons of simplicity, they only depict a single lineage within a group rather than the many lineages that were actually involved. However, these illustrations have often been misunderstood as showing that evolution was working towards a goal.

Missing links

The term 'missing link' is used to describe an as-yet-undiscovered organism that will somehow connect one group with another. The term is most frequently associated with human evolution, and in particular with the idea that there might be a single species that links hominins (modern humans and their close relatives) with non-hominin apes.

The fossil record has confirmed our suspicion that the evolution of hominins and other apes involved numerous species that might be regarded as 'transitional', their bodies combining numerous features typical both of hominins and non-hominins. The idea also implies – wrongly, it turns out – that hominins and apes are separate groups (it's more accurate to imagine hominins as a subset of ape; see page 362). In fact, this theme is typical: because evolution is gradual and involves numerous small steps made across many species within a lineage, there is never any one organism that might be regarded as 'a' or 'the' missing link. Accordingly, the term is generally regarded as misleading.

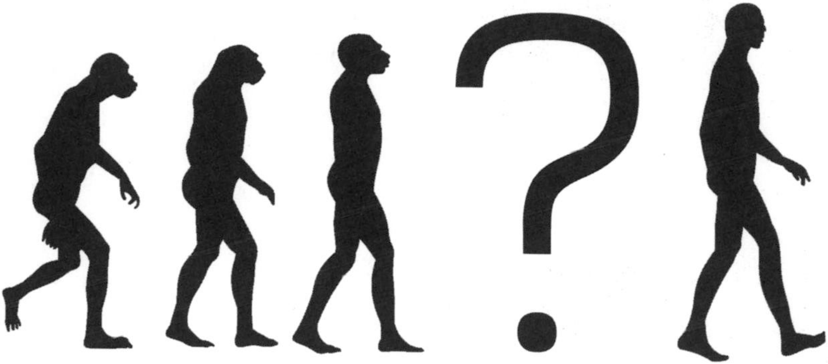

Evolution and ideology

Throughout history, political leaders and parties have used evolution as a means of saying that their specific ideas or messages are the ones most consistent with scientific data. In other words, evolution has been used as a part of the ideology of a given group of people. Given that some, most or many political ideologies have a specific, biased agenda – they seek to establish the superiority of one group of people or nation – it is easy to see how scientific ideas and theories might be twisted and modified in such cases. The Nazis, for example, argued that some groups of people were evolutionarily superior to other groups, and they used pseudoscience to support their claims.

It is often argued by creationists that Darwin's ideas influenced Karl Marx and were thus critical to the communist ideology he devised. But this is not correct. Marx published his key works on communism during the 1840s (*Origin of Species* was published in 1859) and his post-1859 works contain no references to Darwin or to evolutionary theory.

Nazis believed that blond 'Aryans' were superior to other races, and sought to cultivate racial purity.

Nature vs nurture

In complex animals such as humans, the behaviour of an individual as it matures can be regarded as a mix of inherited or inborn traits ('nature') and a set of taught or learned behaviours that are dependent on the environment ('nurture').

It is popularly said that humans (and maybe other animals) are the products of either 'nature' or 'nurture', and a surprisingly common view for much of the 20th century was that humans are almost entirely the product of nurture.

In reality, both of these processes are at work and every organism has to be regarded as a mix of inherited traits and the things that affect it during its development. Consequently, experts today regard the idea of a nature/nurture dichotomy as misleading and the whole idea is mostly ignored. Indeed, the physical characteristics of an organism can be heavily influenced by the environment in ways that are not specifically controlled by its genome (see page 206).

Organisms are the product of their genetics – their 'nature' – but events that occur in their environment – their 'nurture' – play a role in their development too. Studies of twins raised in different environments can reveal some of the factors at play.

Phylogenetic trees and how to build them

Relationships between living things have frequently been depicted in branching diagrams termed 'phylogenetic trees'. During the 1960s, insect specialist Willi Hennig argued that species should only be grouped together on the basis of shared features or 'characters', not on features broadly distributed within the group concerned. Primitive, widespread characters were termed 'plesiomorphies'; advanced characters, specific to a subset of that group, were termed 'autapomorphies'; characters tying branches together were termed 'synapomorphies'.

Originally, people dealt with small numbers of characters and were able to work out for themselves how the distribution of character information could be explained. Their phylogenetic trees, also called 'cladograms', were generally small and simple. As data sets of characters have become larger, computers have become integral to the process. Today, computer programs can analyse huge quantities of character information and generate giant, complex cladograms that best account for the data they analyse.

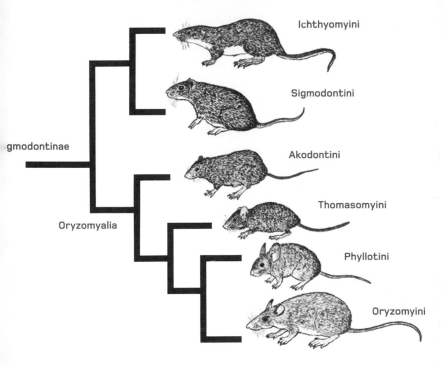

By tabulating data on the different features present in a group of organisms, experts can build phylogenetic trees, or cladograms. This one shows the branching pattern within a group of American mice called the sigmodontines.

Molecular clocks

Genes change over time as a consequence of mutation. It has been proposed that this change occurs constantly, continually, and at a clock-like rate. If the rate at which this change occurs can be determined, the degree of genetic difference between any two species can be calculated with precision. Consequently, the evolutionary split between those species can be pinned down to a specific time in the past. In other words, we can imagine the genome of an organism as a ticking clock that provides actual data on the timing of evolution. This 'molecular clock' hypothesis can also be termed the 'gene clock' or 'evolutionary clock' hypothesis. It emerged during the 1960s.

There have been several criticisms of the molecular clock idea. The notion that changes occur at a clock-like rate is questionable and changes are certainly not constant or similar across different groups of organisms. For these reasons, molecular scientists generally talk today of 'relaxed molecular clocks', where the rate of genetic change varies from one group to the next.

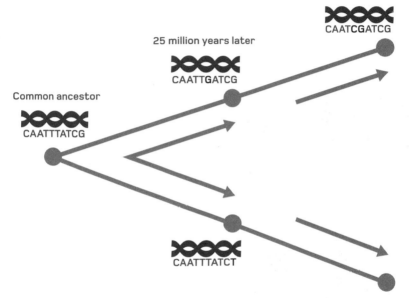

Divergent DNA

50 million years later
CAATCGATCG

25 million years later
CAATTGATCG

Common ancestor
CAATTTATCG

CAATTTATCT

CAATGTATCT

The accumulated differences in the genetic code of two organisms, arising from random mutations, provide an indication of the point of time in the past when the two species diverged.

Cladogenesis

Several different lines of evidence – predominantly fossils and genetics – show that a given species or lineage has sometimes split, resulting in the appearance of two or more species. This process is termed 'cladogenesis', since it involves the creation of a 'clade', this being a group of organisms where all constituent members descend from the same single ancestor. Cladogenesis is typically seen as the opposite of anagenesis (see page 108).

Cladogenetic events might occur for several reasons, the primary one probably involving the moving of an ancestral population into a region where new opportunities are available. Subpopulations of the ancestral one begin to diverge, eventually accruing enough differences that they evolve into separate species. In some cases, so many new opportunities are available that the ancestral population gives rise to a burst of many new lineages, an event known as an 'evolutionary' or 'adaptive radiation'. Some experts argue that the majority of new species arise via cladogenesis.

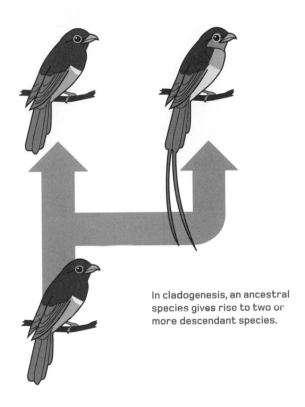

In cladogenesis, an ancestral species gives rise to two or more descendant species.

Anagenesis

Anagenesis describes a model of evolution where a given lineage evolves gradually into another, without any splitting off of other lineages or diversification events. In effect, it describes 'straight-line' evolution in which a species gives rise to a single descendant. Because the pattern concerned is gradual and incremental, species boundaries are somewhat arbitrary. Some experts propose chopping up the lineages concerned into segments variously termed 'palaeospecies', 'chronospecies' or 'successional species'. There are several fossil groups where this pattern is said to be present, including dinosaurs and hominins. Anagenesis is generally regarded as being inconsistent with models of evolution that posit brief bursts of evolution – most notably punctuated equilibrium (see page 134).

The term 'anagenesis' has been used in several ways over the years. When first coined in the 1920s it was specifically used to describe patterns of increasing complexity and was not at all regarded as being incompatible with the process of cladogenesis.

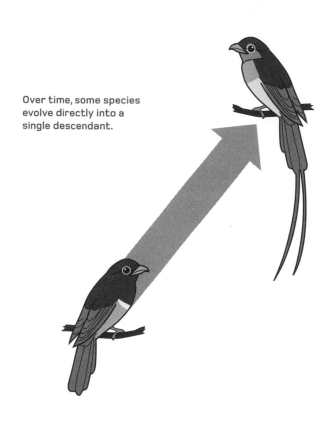

Over time, some species evolve directly into a single descendant.

Convergent evolution

If evolution results in organisms becoming better adapted to a given lifestyle or environment, it follows that similar solutions to similar lifestyles or environments will occur on more than one occasion. Living and fossil animals show us that exactly this has happened numerous times, the process being termed 'convergent evolution'. Simply put, the process refers to the way in which distantly related organisms end up being somehow alike due to their becoming adapted for the same way of life.

A good example of convergence lies in similarities between sharks, ichthyosaurs and dolphins. All three evolved from differently shaped ancestors, but have converged on the same basic body shape. The strong similarity between members of the dog family and the fox-shaped thylacine – a marsupial – is another example. As well as acting on an organism's entire structure, convergence can also affect specific organs, or even refer to pieces of the genetic code or an aspect of behaviour or physiology. Convergent evolution has proved pervasive across the tree of life.

Ichthyosaurs (a group of extinct swimming reptiles) and dolphins (a group of swimming mammals) represent one of the best examples of convergent evolution.

Ichthyosaurs were streamlined and gave birth to live babies in the water.

Dolphins are also streamlined and give birth to live babies in the water.

Evolutionary radiations

For most of geological time, evolution occurs in a generally mundane manner, one species gradually evolving into another or giving rise to one or two new branches that, in time, become new species too. Occasionally, however, unusual circumstances arise whereby a species has numerous opportunities to adapt to several or many new niches simultaneously (see page 68). Different subpopulations of the original species eventually become distinct species as they adapt to different food sources and ways of life, this adaptation virtually always being linked with changes in shape. The result is an explosive pattern where a single ancestor gives rise to many descendant species.

Such events are termed 'evolutionary radiations'. They are a common feature of evolution on islands where an invading species is instantly presented with numerous unexploited niches. They can also occur in the wake of extinction events. Good examples of evolutionary radiations include those discovered in island-dwelling birds, such as the Galápagos finches.

There are many occasions in the history of life when an ancestral species has given rise to a burst of descendants, as shown here in Hawaiian honeycreepers. Such events are termed evolutionary radiations.

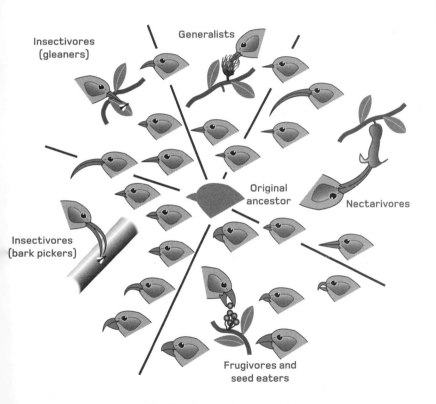

Generalists

Insectivores (gleaners)

Nectarivores

Original ancestor

Insectivores (bark pickers)

Frugivores and seed eaters

Heterochrony

Different parts of an organism's structure develop at different rates. In humans, for example, the limbs increase in size at a much faster rate than the skull does. During the course of evolution, organisms of many kinds have altered the rate at which these sorts of developmental changes occur, as well as the sequence and timing of such occurrences. The whole topic – the timing and nature of developmental change across evolution – is termed 'heterochrony'. The concept originated with Ernst Haeckel during the 1870s.

Several examples of heterochrony are obvious. Among the most famous is the process known as 'paedomorphosis' where features ordinarily only associated with juveniles are retained into adulthood. Amphibians – such as the axolotl – where adults retain the large external gills otherwise typical of juveniles, are good examples of this. Heterochrony is common across animal groups. Many features that make humans unusual compared to other apes, for example, seem explained by it.

Some amphibians, such as this axolotl, have delayed certain parts of their development while other parts have continued to develop at the normal rate.

Thompson and morphogenesis

It is obvious that organisms vary in shape, and that shape is fundamental to an organism's way of life. While the pressures of evolutionary selection have modified the shapes of organisms over time, it is evident that mathematical and physical principles control which shapes are possible and which are not. It took pioneering work by Scottish biologist and mathematician D'Arcy Thompson to bring this concept to widespread acceptance. Establishing the concept of 'morphogenesis', he proposed the idea that factors controlling the shape of an organism are fundamental to its biology. Thompson's ideas were most famously put forward in his 1917 book *On Growth and Form*.

Thompson showed, via the use of grids placed on top of diagrams of animals, that skewing of the grid in one direction would result in the sort of change that would allow the emergence of a new species. These ideas can be considered linked to heterochrony and evo-devo (see page 220).

Morphogenesis

Some organisms (like the sunfish shown right) have seemingly evolved their unusual body shape by modifying the form present in an ancestor.

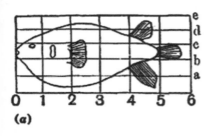

(a)

Ancestral fish

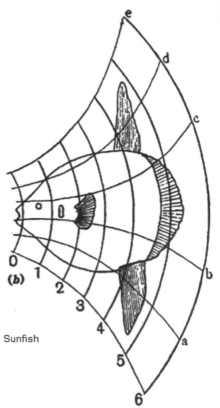

(b)

Sunfish

Hybridization

Until recently it was thought that hybridization between species and lineages was rare or non-existent. Increasingly, however, DNA analysis has shown that hybridization is normal and frequent in many organisms. Some species have arisen as the products of hybridization between two long-distinct lineages. An example among mammals concerns Père David's deer, a weird-looking deer that seems to be a self-perpetuating hybrid between members of the red deer and Eld's deer lineages. Numerous plants regarded as distinct species have also arisen from hybridization.

Hybridization where hybrids between two species then cross back with either of the parental species – an event known as 'introgression' – is also common and has been documented in oaks, sunflowers, irises and flies. These findings show that both species and lineages are 'leakier' than we used to think, that genes can be exchanged across parts of the tree of life, and that hybridization can improve fitness in cases, not decrease it.

It is often said that species cannot produce fertile hybrids. This is not always true. Among ligers (lion–tiger hybrids), the females are fertile.

Hybrid zones

Species are typically understood to be populations of organisms that do not interbreed with other populations also considered to be species. This definition is not workable in all cases, however, since successful hybridization is frequent (see page 118). Indeed some species hybridize so frequently that the areas in which their ranges overlap are permanently occupied by hybrids of the two. Such zones are termed 'hybrid zones'.

Hybrid zones are important for our understanding of evolution because they indicate that the two species involved have not been separated for long and are still exchanging genes. They are typically linked with 'parapatric speciation', a form of speciation (see page 124) where members of a population move into a new area and begin to evolve into a new species. Some hybrid zones might show that the parent species have only recently come back into contact, having previously evolved in isolation for some period of time. The best known hybrid zone is the European one populated by hybrids of the hooded crow and carrion crow.

The carrion and hooded crows are two separate species inhabiting eastern and western Europe respectively. Where their habitats overlap, however, the two mate and are almost identical, genetically.

The web of life

We've already seen how the relationships between living things — the overall pattern that has given rise to the diversity of life — can be imagined as an enormous, phenomenally complex tree. Branches are generally regarded as having their own 'destiny', forever moving away from the trunk of the tree once they've emerged as a distinct unit. However, recent discoveries have shown that this is somewhat oversimplified, since it turns out that many branches are not as distinct as we might think. For example, organisms that have been separate for a long time can still exchange genes. This can occur via hybridization or via the horizontal gene transfer common across plants or that occurs when parasites invade their hosts (see page 216).

A consequence of these connections is that it might be more accurate to refer to a 'web of life' more than a tree of life, and this term has become more common in discussions of phylogeny since about 2008. The term has also been used in relation to the connections present across ecological communities.

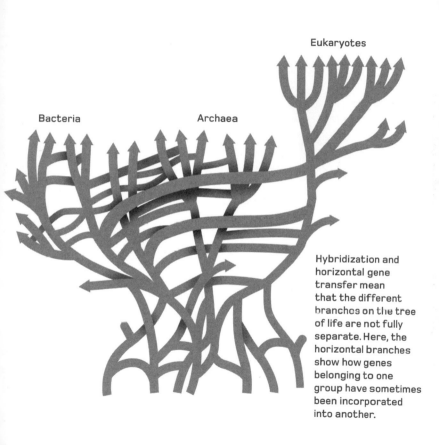

Eukaryotes

Bacteria

Archaea

Hybridization and horizontal gene transfer mean that the different branches on the tree of life are not fully separate. Here, the horizontal branches show how genes belonging to one group have sometimes been incorporated into another.

The process of speciation

How, exactly, do new species evolve? This issue – termed 'speciation', the core question of Darwin's 1859 book *Origin of Species* – is one of the most discussed and most controversial areas in evolutionary theory. Darwin, and others of his time, did not have a clear model on how species actually arose. Tackling this question requires that experts agree on what species are in the first place, and it took some time before any consensus was achieved on this issue (and it remains hotly debated even today).

During the middle of the 20th century it became increasingly obvious that species mostly arise as a consequence of isolation, a view promoted in particular by evolutionary biologist Ernst Mayr. However, gene exchange across a population can mean that large sections of that population can evolve in the 'same direction' all at once, the result being that one species might evolve directly into another one. Other mechanisms allowing speciation also exist (see page 126).

Some organisms have only recently evolved from an ancestral species, and in some cases are still in the process of emerging as a distinct species. Several recently emerged species are present in the white-headed gulls, the group that includes the herring and lesser black-backed gulls.

Allopatric and sympatric speciation

There are several different ways in which new species might arise. Among the most important are allopatric and sympatric speciation. Both were made well known by pioneering evolutionary scientist Ernst Mayr in his writings of the 1940s.

'Allopatry' describes a situation in which populations are separated geographically, either because one of the populations has dispersed over some distance or because a barrier – a river, valley or mountain range – has arisen. This separation eventually results in the evolution of enough differences that speciation occurs. Allopatric speciation appears to be common. 'Sympatry' describes a situation in which the populations concerned live alongside one another. Due to some behavioural, ecological or anatomical change, members of a population become distinct and stop breeding with those of the ancestral population, instead forming a new, distinct subpopulation. Eventually, the younger population accrues enough differences to become a distinct species. Confirmed cases of sympatric speciation are rare.

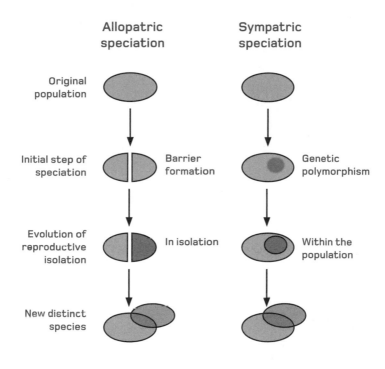

	Allopatric speciation		**Sympatric speciation**
Original population			
Initial step of speciation		Barrier formation	Genetic polymorphism
Evolution of reproductive isolation		In isolation	Within the population
New distinct species			

The ring species model

Imagine a population that occurs along a straight line, and is divided into subpopulations that differ in minor ways. Adjacent subpopulations can interbreed, so there is genetic continuity across the entire population. However, the subpopulations at the extreme ends of the line are distinct enough that they cannot interbreed should they meet. Imagine that line on our spherical planet; the result is a ring-like model where the end points are distinct species, even though they are part of the same continuous unit. This is the 'ring species model'. It was used for several decades as a way of showing how new species can evolve via the accumulation of small differences accruing over distance. Several animals were used as examples of the ring species model, including the herring gull and the Tibetan warbler. Genetic studies have shown that none are ring species after all, since adjacent subpopulations are either not especially closely related, or were separated at various points in the past and hence did not indulge in gene flow. Nevertheless, the model continues to exist as a theoretical idea that might, in principle, be verified.

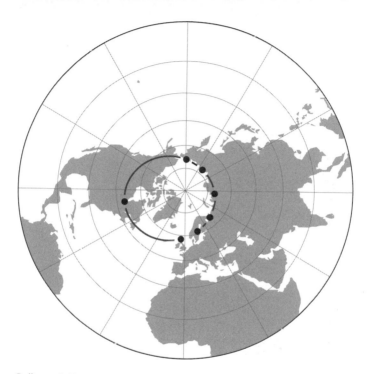

Gull populations arranged in a ring-like pattern across the northern continents were once thought to be genetically continuous 'super-species'. In fact, they are not.

Incipient species

The theory of evolution proposes that species change over time, giving rise to new species. Both theory and observation indicate that such origination events do not concern all members of a species, however. Instead, a population *within* a species becomes isolated in some fashion from others – perhaps because it lives on an island, or takes to a new way of life. It then evolves novel features of its own.

When distinct enough from other members of their species, these populations are typically designated as distinct units termed 'subspecies'. In many cases, these subspecies are literally evolving into distinct species of their own and are what we term 'incipient species'. Numerous examples are known, and sometimes the populations concerned have become more distinct from their ancestral populations in the time that we have studied them. They include island-dwelling mice and birds, fish that are evolving different migratory habits from their ancestors, and flies that are adapting to exploit different fruiting trees.

Like the Eurasian jay, many birds consist of populations that are gradually becoming so distinct that they will surely evolve into new species. Some may be deserving of the species title already.

The species debate

It has always been obvious that living things are sorted into units – species – that look different from the members of other units. It is mostly understood that members of one species look alike, behave in similar ways, interbreed and avoid breeding with those of other species. This view is termed the 'biological species concept' (BSC). It is popular and widely accepted. The problem is that there are many exceptions to the BSC. For example, there are species in which individuals display radical diversity in appearance, and in which individuals successfully hybridize with members of other species.

Consequently, numerous attempts have been made to redefine 'species'. There are currently some 26 such definitions, though many overlap; some only require that a unit we term a species represents its own distinct branch on the tree of life. Furthermore, biologists not only disagree on how species should be defined, but also on which populations should be regarded as species and thus on how many species there are.

Experts disagree as to whether certain organisms—like the red wolf shown here—should be regarded as true species or not.

Punctuated equilibrium

During the 1970s, palaeontologists Niles Eldredge and Stephen Jay Gould rejected the idea that evolution was a slow and gradual process. Instead, they said, the fossil record indicates long periods of stasis followed by brief, tumultuous periods of rapid change. They described evolution as being marked by sudden jumps, and termed their model 'punctuated equilibrium'.

Punctuated equilibrium was widely misunderstood. It does not propose – as stated by some creationists – that major changes occur over one or two generations, but that the main periods of natural selection and evolutionary change are compressed over tens of thousands of years, rather than being continuous during the whole time that a population exists. The idea that evolutionary change occurs at variable speed across the history of a species is now generally accepted. However, it remains debatable whether the periods of change identified as key to punctuated equilibrium actually contradict models of evolution that predict more gradual rates of change.

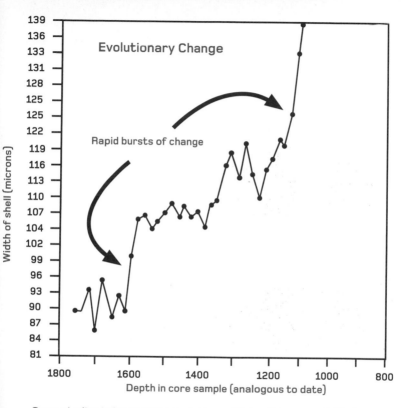

Evolutionary Change

Rapid bursts of change

Width of shell (microns)

Depth in core sample (analogous to date)

Some studies indicate that populations of living things are static for long periods of time then suddenly undergo rapid bursts of change. This chart maps such changes in single-celled marine organisms called radiolarians.

Intelligence and consciousness

How intelligence and consciousness might have evolved remains a heated topic of debate. There is little question that intelligence (an ability to retain information and apply it to new situations) is widespread across animals, being pronounced in primates, parrots, crows and dolphins. The more we learn about other animals, the more widespread intelligence appears to be. Recent studies ascribe intelligence to rodents, monitor lizards and manta rays. As a general rule, large-brained animals with complex social lives exhibit behaviours that we identify as intelligent, and it may be that intelligence has evolved in response to social complexity and a lifestyle in which the application of flexibility and memory are crucial.

We do not yet fully understand how consciousness works, but an idea that is becoming popular is that it evolved as a way in which complex organisms could devote attention to a task by being aware of the fact that the task was being performed.

Microevolution vs macroevolution

It is sometimes argued that there are two kinds of evolution: the kind that involves small changes at the population level, termed 'microevolution', and the kind that involves more substantial change as one group gives rise to another, termed 'macroevolution'. Both terms were first introduced in the 1920s.

Macroevolution is associated with major changes in shape, with the origin of new structures and new ways of life, and with the origin of wholly new groups. The evolution of whales from land-living mammals would be an example of macroevolution. However, the two kinds of evolution are not really different phenomena: when enough small, microevolutionary changes accrue, the result is a change of the sort we conventionally imagine as macroevolution. In other words, major, swift evolutionary events – those involving gene duplication, for example – would be considered macroevolutionary. But this kind of evolution might just as readily result from the accruing of a large number of small changes.

Small-scale evolutionary events, such as the evolution of one warbler species into another, are examples of 'microevolution'. Longer term transitions that involve a greater degree of change, such as that involved in the evolution of whales, are examples of 'macroevolution'.

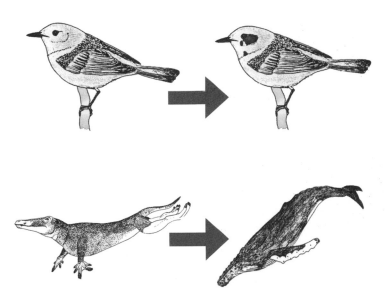

Developmental plasticity

Variation within species is a key aspect of evolution. This is particularly true of species that vary a huge amount according to the conditions of their growth. Examples include the ability of some amphibians to develop a fully aquatic lifestyle when conditions require it; the development in some waterfleas of larger, spinier individuals when predators are present; and the tendency of some amphibious plants (like water buttercups) to grow different leaf shapes depending on whether the leaves are submerged or kept above water. Organisms that possess this ability to be highly variable are described as exhibiting 'developmental plasticity'. It may mean that the organisms concerned are especially resistant to selective pressures and therefore especially good at persisting for long periods of time. However, the fact that some 'plastic' organisms tend to generate entirely new biological structures or behaviours also means that they are especially likely to give rise to new forms when local conditions give them an advantage. Plasticity may therefore be key in the evolution of new species and even whole new groups.

Oystercatchers have highly variable bill shapes. Some individuals (left) have knife-like bills used to slice open shells, while others (right) have club-like bills better suited to smashing shells open.

Optimal and bad design

Research has shown how many organisms exhibit 'optimal design', this meaning that they are as well shaped or proportioned as they can be to perform the tasks they do. In other words, adaptation and natural selection have fine-tuned organisms in ways that exceed or parallel those of the best human designers. Examples include the tubular jaws of pipefishes and the shells of some turtles, which are ideally shaped and weighted to allow the animals to self-right after toppling over. However, many organisms appear 'badly designed' for certain behaviours. For example, humans suffer numerous spinal problems from walking upright.

It could be that those organisms with optimal design have been adapting to a given way of life for a long time, or have adapted to a way of life for which biological structures can easily become shaped. Conversely, bad design may be present because organisms have not been adapting for the given behaviour for an especially long time, or are not yet specialized for it.

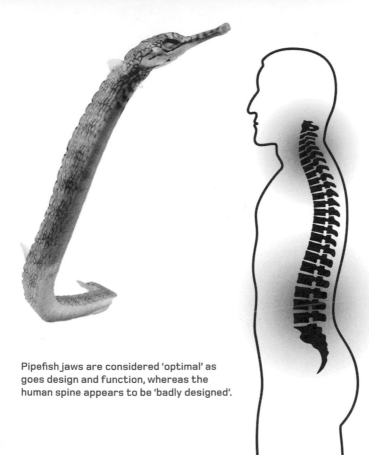

Pipefish jaws are considered 'optimal' as goes design and function, whereas the human spine appears to be 'badly designed'.

Constraints
and limitations

The process of evolution means that organisms can adapt to new situations, whether they be ways of life or environments. However, organisms cannot evolve infinitely, or adapt to any given situation: there are numerous constraints or limitations that prevent certain things from happening. These factors are important in evolution and explain why many particular events have not occurred, and why certain niches have remained unfilled.

Size is one particular limitation, the anatomical configuration of a given group meaning that it cannot evolve above or below a given size. Vertebrates, for example, can only grow so big before bone and cartilage exceed their ability to safely carry weight, and blood vessels can only become so small before it becomes difficult for them to transport liquids. Numerous constraints control shape in organisms. Patterns of development mean that organisms simply have to have their structures arranged in a certain order (see page 94), and deviations from these orders are usually lethal in developing embryos.

Giant organisms face structural challenges that put limits on their size. The biggest of these is gravity, which is why whales, free of its constraints, can grow signficantly larger than land animals.

Biomechanics and evolution

Structures in biology have to make sense in engineering terms. A plant stem or animal limb must be able to support the weight of the rest of the organism; jaws and teeth have to be strong enough to break and crush food items. If the basic mechanical demands of a structure are not met, the organism that possesses them will be unfit and will fail to produce surviving offspring successfully. The study of how biological structures function and perform in mechanical terms is called 'biomechanics'.

Some structures in biology are not optimally designed (see page 142). However, it is generally assumed that the majority of structures that have evolved in organisms operate at close to a biomechanical optimum. Particularly famous studies have examined the performance of mammal jaws in chewing, the digging styles used by burrowing reptiles and feeding behaviour in snakes. Increasingly, engineers look at naturally evolved structures to see how devices and objects made by humans can take advantage of designs already invented by evolution.

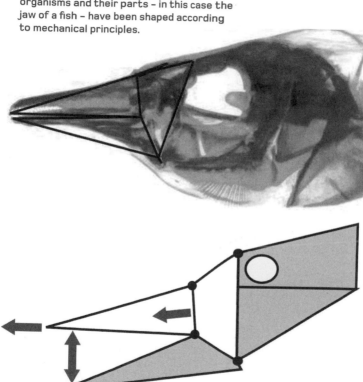

By examining biological structures as mechanical units, biologists can show how organisms and their parts – in this case the jaw of a fish – have been shaped according to mechanical principles.

Sociobiology

When we discuss the evolution of traits, we normally have physical components of anatomy or segments of the genetic code in mind. However, behaviour is also subject to evolution. In his 1975 book *Sociobiology: The New Synthesis*, E. O. Wilson argued that social behaviour can be seen as the outcome of evolutionary mechanisms. The resulting field of study is termed 'sociobiology'. Wilson argued that aggression, parental behaviour and herding behaviour, for example, are all passed down genetically because they are adaptive: in the animals that possess them, they are now intuitive, instinctive or inborn. This can even apply to altruism (see page 218).

Some contest sociobiology, arguing that social behaviour is not solely explained by inheritance, but by learning and environment, too (a debate identical to the 'nature vs nurture' one). The idea that social behaviour is genetically predetermined has also been controversial in discussions of human aggression since it could be taken to show that aggression is inevitable.

Sociobiology explains cross-species friendships such as animal adoption in terms of a response to stimulus—in this case, the cat's nurturing instinct has overcome its predatory one.

The island rule

Organisms that adapt to life on islands evolve under a different set of selection pressures than those living on continental landmasses. Food and other resources are typically more restricted on islands. Predators may also be reduced or absent altogether, and competition from other species less severe. Owing to such different selection pressures, island-dwellers may be more variable and more unusual in form than their continental relatives. For example, reduced resources may lead to some large-bodied animals becoming smaller over time; and reduced or absent predation may result in other species becoming larger than their continental relatives. The entire phenomenon of unusual evolution on islands is typically called the 'island rule'.

The island rule is most extensively studied in mammals, tortoises, lizards and birds. Classic examples include island-dwelling giant tortoises and the recently extinct dwarf elephants and giant dormice of various Mediterranean islands.

Group selection

It is typically thought that evolution works at the level of the individual – that individuals are selected through natural processes to pass on their traits, and that individuals compete with one another for resources or mates. Observations indicate, however, that evolution sometimes involves cooperation between individuals, the members of a group somehow evolving in step. This model is termed 'group selection'. Its existence, and the mechanisms behind it, form one of the most controversial areas of evolutionary theory.

A classic example involves social insects: animals such as termites, wasps, ants and bees that encourage the survival of entire colonies over the interests of individuals. In such cases, evolution can only have happened through cooperation between kin. Other traits present across whole populations of animals (such as restraint in reproduction) are also argued to represent group selection, since they encourage survival of a whole colony over the selfish interests of individuals.

Metabolism and evolution

Evolution does not occur at the same rate in all species, or even in all individuals of a species. This is confirmed by observations of molecular change in living organisms. There are several reasons for variable evolutionary rates. Some organisms have a genetic chemistry that results in slow DNA replication and repair. Viruses, on the other hand, have especially high rates of DNA change, perhaps because a rapidly changing structure makes it difficult for hosts to evolve immune systems capable of defeating them.

A key factor controlling the rate of change in many organisms is metabolism: the rate at which chemical processes occur within the body. Some of the best illustrations of how metabolism affects the evolutionary rate come from studies of mammals. Rodents undergo genetic changes more rapidly than primates do, and some primates have undergone a marked slowing of molecular change. These different rates of change are linked to metabolism and hence to diet and other key aspects of biological adaptation.

It is generally thought that high metabolism – like that present in a hummingbird – results in a faster rate of genetic change and thus faster evolution. Conversely tortoises have slow metabolisms alongside a more measured pace of evolution.

Gigantism and dwarfism

Over time, some organisms have increased markedly in size relative to the condition of their ancestors. This trend is termed 'gigantism'. In opposition, some organisms have reduced markedly in size relative to their ancestors, a trend termed 'dwarfism'. Gigantism is illustrated by such animals as sauropod dinosaurs and baleen whales, but also by numerous other living things, including molluscs, trees and kelp. Dwarfism is best known in frogs and lizards where there are tiny species less than 2 cm (¾ in) long, but has also occurred in the evolutionary history of some fishes, most notably sticklebacks and gobies. Both gigantism and dwarfism are at play on islands (see page 150).

There are several reasons why members of a lineage might become gigantic over time; it may arise from an evolutionary response to an abundance of resources, to the avoidance of predation, or as a way of out-competing other species. Dwarfism may be a consequence of reduced resources or may represent adaptation to a niche in which small size has been selected.

Selection pressures make some species small and other species large. Chameleons include both dwarfs and giants.

The handicap principle

Both natural selection and sexual selection can encourage the evolution of elaborate, complex structures in organisms. Sexual selection (see page 66) is unusual in respect to natural selection in that the development of structures that prove detrimental to day-to-day survival are encouraged. Examples include deer antlers, large showy display feathers on birds and eye stalks and elaborately patterned wings on some flies. Some of these structures are so large, complex and visually obvious that they have a significant survival cost for the organisms that bear them: they provide a 'handicap' to survival. Because they are so difficult to survive with, these structures (when in good condition) demonstrate that the organism bearing them is strong, healthy and a good mate. In other words, the structures are honest indicators of genetic qualities that are attractive to potential mates. It is for this reason that such structures persist, and even become larger and more elaborate over time. The phenomenon driving the evolution of these structures is termed the 'handicap principle', first proposed by Israeli biologist Amotz Zahavi in 1975.

Several bird groups have evolved large, showy ornaments that have an impact on their survival. Hornbills are one such group.

Runaway selection

Sexual selection promotes the evolution of gaudy, elaborate structures that are advantageous because they increase mating success. If these structures do indeed signal fitness, there will be evolutionary tendencies for the choosy gender (typically the females) to become ever choosier over time, and for the elaborate gender (typically the males) to become more elaborate over time. This link is known as a 'positive feedback'; if it continues for long enough it will lead to 'runaway selection', the evolutionary process whereby elaborate structures are driven to become ever larger, ever more elaborate.

Structures that exhibit runaway selection sometimes pose a significant survival cost to their bearers and thus some are also linked with the handicap principle. Birds with especially elaborate plumage – typically peacocks and widowbirds – have most frequently been regarded as examples of runaway selection. Others include certain water boatmen, a group of aquatic bugs whose loud calls seem to result from the phenomenon.

Selection for extravagant structures can lead to ever larger, ever costlier structures used in sexual display. The gigantic antlers of the extinct deer *Megaloceros* are a classic example.

Evolution and escalation

The process of adaptation (see page 60) means that organisms are constantly having to change to keep in step with others. For example, if a predator needs to continue preying on a given prey species, the two have to change in step if the predator is to persist. This idea is key to the 'red queen' hypothesis proposed by Leigh Van Valen (see page 168). The consequence is that, over time, adaptations related to defence and predator avoidance become more complex. Adaptations to thwart defence and subdue prey species become more complex too. This idea of increasing complexity, of predators and prey evolving adaptations in step, is termed the 'hypothesis of escalation'.

One of the key advocates of the escalation idea is biologist Geerat J. Vermeij. Vermeij noted that the shells of Pacific molluscs tend to be thicker, more complex and more ornate in places where these molluscs had co-evolved with predators adapted to break open shells.

Exaptation

It is typically thought that anatomical structures in animals, plants and other organisms originate following adaptation to a given environment or lifestyle. However, it also seems to be the case that some structures, behaviours and other traits represent modified versions of pre-existing forms that played a different role in a previous mode of life. Feathers, to take one example, are key to a bird's flight ability, but they seem originally to have evolved to provide insulation. The changing role of a trait inherited across evolution is termed 'exaptation', and was first proposed by Stephen Jay Gould and Elisabeth Vrba in 1982. An older (and less preferred) term 'pre-adaptation' means the same thing.

Many structures and other traits that have proved significant in evolutionary history may have originated as exaptations. In vertebrate history, limbs and digits may originally have evolved to assist locomotion in cluttered aquatic environments and only later proved integral to movement on land, for example.

Feathers originated in ground-dwelling dinosaurs and were originally nothing to do with flight. Only later did feathers become exapted for such a role.

Modular evolution

It is generally believed that evolutionary changes occur across the entire structure of an organism at an approximately continual rate. However, the bodies of at least some animals show that this is not always the case. Instead, given sections of the body – termed 'modules' – evolve at slightly different rates according to how selection pressure is acting on those modules. Take birds: the wing module may operate under selection that does not affect the head-neck module or the hindlimb module in the same way or at the same speed. The wing module might change more drastically and more quickly, therefore. This non-simultaneous change is termed 'modular evolution'.

It is probably not coincidental that flying vertebrates are among the best examples of modular evolution, since different modules of the body have become specialized for new functions. The Jurassic pterosaur *Darwinopterus* from China is a classic example: the head-neck module appears significantly more specialized than the archaic forelimb and hindlimb modules.

The Jurassic pterosaur *Darwinopterus* is a classic example of modular evolution: it is almost as if different parts of its body evolved at different speeds.

The red queen hypothesis

The 'red queen hypothesis' proposes that organisms must continually adapt in order to avoid extinction: they cannot remain static and unchanging, since they exist in a world in which all other organisms are evolving as well. The hypothesis points to the existence of an 'arms race' in which organisms are continually forced to keep improving and updating their adaptations. The hypothesis was first proposed by palaeontologist Leigh Van Valen in 1973. Its name is based on the behaviour of the red queen in the 1871 book *Through the Looking-Glass* by Lewis Carroll; she describes how one must run as fast as possible in order to remain in the same place.

The hypothesis is conceptually similar to ideas about evolutionary escalation. It also has relevance to the concept of so-called living fossils (see page 174), since it effectively states that they cannot truly exist given the need for continual change. Work on living fossils is in agreement with the concept, since none of the so-called living fossils have really been static in evolutionary terms.

Atavism

Throughout history, people have recorded unusual cases where young animals (most typically humans) have been born with a structure not typical for their species. Classic examples include external tails on humans, side toes on horses and hindlimbs on whales. These features obviously hark back to an earlier time in the evolutionary history of the lineage concerned: humans descend from primates that once had tails, ancient horses were multi-toed and the ancestors of whales possessed hindlimbs. These one-off recurrences are termed 'atavisms', sometimes popularly termed 'evolutionary throwbacks'.

Atavisms show that the information required to generate structures that existed deep in the history of a given lineage are still present within the organism's genetic code, but are not normally expressed in the live organism. While atavisms do not typically prove advantageous and seem unlikely to persist, there are cases where they might provide an adaptive advantage and hence might become a normal feature once more.

There are many recorded cases of modern whales and dolphins in possession of hindlimbs. These are an atavism, revealing that the genes allowing expression of hindlimbs are still present in these animals.

Wholesale taxic atavism

The idea that certain sections of an organism's body represent atavisms is a familiar one (see page 170). Less familiar is the idea that the entire structure of an organism might be atavistic – that is, the whole of its body has reverted to something that resembles an earlier stage in evolutionary history. This phenomenon is called 'wholesale taxic atavism' or 'whole taxon atavism' (the term 'taxon' can refer to a species or to any group containing one or more species).

A classic example is the gharial, or gavial, a long-snouted crocodylian from southern Asia. The overall anatomy of the gharial indicates that it is ancient compared to crocodiles and alligators, and that its lineage evolved before the common ancestor of crocodiles and alligators was in existence. However, DNA evidence contradicts this: it seems to show that the gharial is not ancient at all, but actually a kind of crocodile. If this is true, the gharial shows wholesale taxic atavism.

The gharial of tropical India is one of the world's most remarkable crocodylians. Its evolutionary history is remarkable too.

Living fossils

A huge number of living things closely resemble ancient relatives known from the fossil record. These organisms seem to have changed little – or even not at all – and consequently they are popularly termed 'living fossils'. Classic examples include the shelled mollusc nautilus; the shelled, burrow-dwelling lingula; coelacanths; and the lizard-like tuatara of New Zealand. The fact that some organisms have changed so little, whereas others have changed drastically over the same time period, is typically explained as the result of adaptation to a lifestyle or environment where there is little or no selection. Living fossils are even used by some creationists as support for their view that evolution does not occur.

It turns out, however, that none of the organisms imagined as living fossils are 'non-evolvers'. Rather, they are slow evolvers, and the majority are, in fact, notably distinct from their fossil relatives. Indeed the term 'living fossil' is so misleading that some scientists argue that it should be abolished.

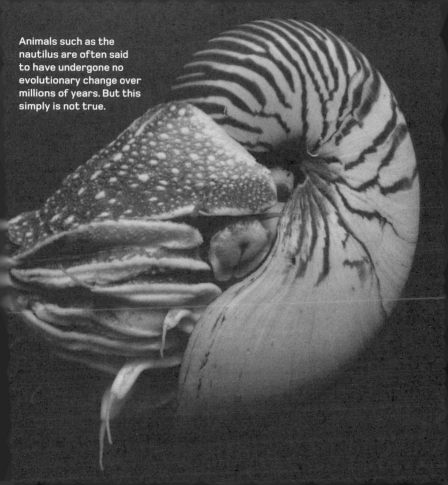

Animals such as the nautilus are often said to have undergone no evolutionary change over millions of years. But this simply is not true.

Evolutionary reversals

It is popularly supposed that features lost during evolution cannot reappear in the descendants of those organisms that lost them. This has even been considered a rule, termed 'Dollo's law'. Both the fossil record and the anatomy of living organisms show that this is not accurate and that 'lost' structures can be regained in evolutionary events known as 'reversals'.

Studies that examine the distribution of traits across groups indicate that reversals are common. A prominent example concerns the lower-jaw teeth of the frog *Gastrotheca*. No other living frog possesses lower-jaw teeth and *Gastrotheca* is not a remnant of an early phase in frog evolution. A reversal is the only process than can explain the presence of its teeth.

In some cases, reversals represent the evolution of a structure or condition that merely resembles that present earlier in evolutionary history. In other cases, however, they literally are reversals, since they involve formerly dormant genes.

The South American hoatzin is often mentioned as a possible evolutionary reversal—large claws on the wings of its babies seemingly reflect an earlier phase in evolution.

Cope's rule and body size

The discovery of numerous fossil mammals during the last decades of the 1800s made it clear that mammals of many sorts had become larger during the course of their evolution. The most famous example concerns horses, but the trend is true for other groups as well. This tendency towards increasing size was identified by American palaeontologist Edward Cope during the 1880s and is generally known as 'Cope's rule'. Cope was far from the only expert to bring attention to the 'rule' and it has also been termed the 'Cope-Depéret rule' to note its promotion by French palaeontologist Charles Depéret.

Experts have disagreed as to how much of a rule Cope's rule actually is. There is little question that body size has increased in many groups of organisms, but this is probably best imagined as a general tendency, not a strict, invariable rule. Indeed, some groups of organisms do not reveal any distinct directional trend relating to body size, while others have become smaller over time.

Groups of organisms have a tendency to become larger over time. This tendency is especially obvious in vertebrates such as dinosaurs.

Early Jurassic sauropod

Mid Jurassic sauropod

Late Jurassic sauropod

Evolution and climate

Climate is one of many factors that influence evolution in organisms. Others include predation, the need to pass on DNA, and the need to find and consume food. Different organisms responded to climate in many different ways. In general, features associated with heat retention are common in organisms that evolve in cool or cold places, and features associated with heat dissipation are common in those of warm or hot places. Warm climates allow some organisms to grow quickly and more continually year-round than they can in cold climates (given the reduced effect of seasonality).

Climate has several other impacts that make the situation more complex, however. Cold water is better at storing oxygen than warm water, which means that aquatic organisms in cold places can be faster and more active; they can forage more quickly and move greater distances. As a consequence, aquatic environments of the polar regions are thus – in places – richer and with a greater diversity of living things than warm aquatic environments.

Dispersal

The distribution of organisms around the world is key to evolutionary history, and the means by which organisms have spread is linked to their origins and adaptations. Many organisms have spread via 'dispersal', the process of moving across distance, either by migrating over land, by air if they can fly, or by water if they are good at floating or swimming. Plant seeds and parasites can disperse within, or attached to, larger organisms. Organisms also disperse across water via the use of floating masses of plant material, a technique termed 'rafting'. Dispersing to a new area is a driver of allopatric speciation (see page 124).

Given the distances that some organisms travel within their lifetimes, and the enormity of geological time, it is unsurprising that living things of many sorts have dispersed across huge distances. For all these powers of dispersal, however, there are still barriers that sometimes prevent spread. These include deep marine channels that are not crossed by convenient currents, and substantial mountain ranges, river systems and deserts.

Many plants have evolved feathery seed-carrying structures to catch the wind and aid the dispersal of offspring.

Vicariance

The process of dispersal means that organisms move to new areas where, separated from their ancestors and subjected to new selection pressures, they typically evolve (via allopatric speciation, see page 126) into new species. But dispersal is not the only way for organisms to become distributed across distance. Neither is it the only way in which allopatry results in speciation. 'Vicariance' is the process by which organisms become separated from their ancestors via the appearance of a barrier, such as a new mountain range, river or sea. Because the original population is divided, the constituent parts evolve in different directions, eventually giving rise to new species.

Because vicariance results from the appearance of a barrier, any group of organisms affected by a vicariant event must be geologically younger than the barrier itself. While vicariance has often been seen as the 'opposite' explanation of distribution relative to dispersal, it is important to note that both processes can operate across the history of a given group of living things.

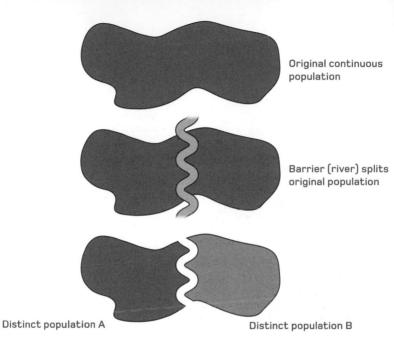

Original continuous
population

Barrier (river) splits
original population

Distinct population A

Distinct population B

A population that occurs continuously across a region might, at some
point, be split in two by a barrier, such as a river. The inevitable result
of this split is the evolution of separate, distinct populations that
eventually become two species.

Allen's, Bergmann's and Gloger's rules

The demands of living at particular latitudes, with specific light levels and in certain temperatures, have led to several somewhat predictable trends in the evolution of living things, especially in animals, and to a large extent in mammals. A number of these trends have been declared as 'rules' or 'laws' that link given details of anatomy with adaptation to distinct conditions.

'Allen's rule' argues that the extremities of warm-blooded animals – limbs, ears and tails – tend to be short in cool climates and large in warm ones, owing to the need to either prevent or promote heat loss. 'Bergmann's rule' proposes that warm-blooded animals become larger closer to the poles, since a greater volume results in a smaller surface area and improved heat retention. 'Gloger's rule' has it that animals in low latitudes tend to be more heavily pigmented than those from higher latitudes.

The validity of some of these rules – Bergmann's rule, in particular – has been questioned.

Clines and environmental gradients

Virtually all organisms that occur within any span of distance vary across that distance, typically in a gradual way such that individuals at the extreme ends of a range are connected by numerous intermediates. This can be imagined as the numerous shades of grey that connect the extremes of black and white. If, as an example, we imagine an organism that occurs in a cold climate at one end of its range, and in a warm climate at the other, we see cold-adapted and warm-adapted individuals at the extremes, linked in the middle by a range of intermediates.

These ranges of continuous variation are termed 'clines' or 'evolutionary gradients' and they show how a species has adapted to local conditions across its range. Clines are present in organisms that have adapted to local climates, and also in those that have adapted to different elevations, light levels and genetic factors. Ring species (see page 128) were once regarded as clines but are probably not. Clear clines are present in many large mammals and widespread bird species.

Many species vary across their range due to environmental variation. Brown bears in Asia, for example, become larger and darker in the north and east of their range.

Neo-Darwinism and the modern synthesis

Darwin's theory of evolution involved both descent with modification *and* the phenomenon of natural selection, the latter acting upon variation and thus acting as the main mechanism of evolutionary change. However, the exact way in which information was passed down the generations remained opaque. During the 1930s and 1940s, several biologists realized how the genetic experiments conducted by Gregor Mendel (see page 48) provided the answer: information was transmitted by genes, and changes in the frequency of alleles (gene variants; see page 206) explained the variation seen in any set of offspring.

This merging of genetics with Darwin's view of evolution by natural selection was termed the 'modern synthesis', and the scientists behind its inception were Theodosius Dobzhansky, Ernst Mayr, G. Ledyard Stebbins and George G. Simpson. This new way of looking at evolution became known as 'neo-Darwinism'. It seems weird to think that evolution might be considered *without* genetics at its core, but it took time to merge these fields together.

Genetic studies – often carried out on laboratory organisms such as fruit flies – have become increasingly integrated with discussions of the fossil and anatomical evidence for evolution.

Phylogenetic systematics

The study of how organisms are related to one another – how they are arranged on the tree of life – is termed 'phylogenetics'. Traditionally, organisms were often grouped together on the basis of general similarity and people were somewhat vague as to which traits allowed them to determine the affinities between species. During the 1960s, the entomologist Willi Hennig proposed that organisms should only be grouped together when it could be shown that they descended from a single ancestor, and when there were specific traits (termed 'characters') tying them together and distinguishing them from other members of their group (see page 102).

A group united by specific traits, and thought to have descended from a single common ancestor, was termed a 'clade'. Hennig's more scientifically rigorous practice of phylogenetics became known as 'phylogenetic systematics'. But since his philosophy was built around the discovery of, and relationships between, clades, the science soon became known as 'cladistics'.

Hennig and his colleagues used anatomical details of insect bodies—like those concerning wing venation—to work out how these animals might have evolved.

Prokaryotic cells

In the simplest terms, living organisms fall into two enormous groups: those that consist of single cells that are simple overall and those that consist of numerous cells, many of which are complex. Single-celled, simple organisms are called 'prokaryotes'. They lack the nucleus, mitochondria and other components typical of eukaryotes (see page 196) and hence have generally been considered more primitive or ancestral.

Traditionally, prokaryotes were divided into two groups: bacteria and archaea. However, genetic studies show that this classification is too simple and that prokaryotes actually include numerous distinct evolutionary branches. Prokaryotes encompass an extraordinary diversity, exceeding eukaryotes as far as lifestyle and internal chemistry are concerned. They include thousands of planktonic species; extremophiles that inhabit hot springs and deep-sea vents; symbiotic organisms that live inside other things; and subterranean species that break down inorganic chemical compounds.

Prokaryotic cell

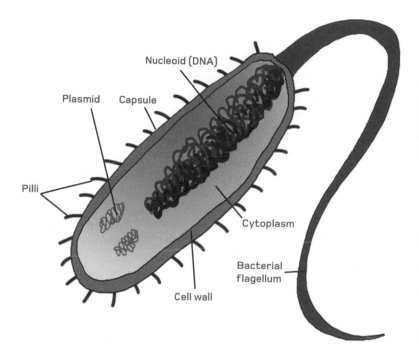

Nucleoid (DNA)

Plasmid

Capsule

Pilli

Cytoplasm

Cell wall

Bacterial flagellum

Eukaryotic cells

Eukaryotes are those organisms consisting of cells in which there is a distinct nucleus, a mitochondrion and other distinct internal structures, termed 'organelles'. Photosynthesizing eukaryotes also contain the packages of green pigment termed 'chloroplasts'.

Because eukaryotes are generally larger and more structurally complex than prokaryotes (see page 194), it is widely thought that they descend from prokaryotes. In some respects, eukaryotes are similar to bacteria, and in others they resemble members of the prokaryote group archaea (see page 230). The very first eukaryotic cell was presumably the result of an event where several prokaryotes became united in symbiosis (see page 232). Eukaryotes include an enormous diversity of single-celled microscopic organisms – grouped together as 'protists' – as well as multicellular organisms, namely plants, fungi and animals. In contrast to prokaryotes, eukaryotes appear to be a clade, and one that originated around 1.6 to 2.1 billion years ago (bya).

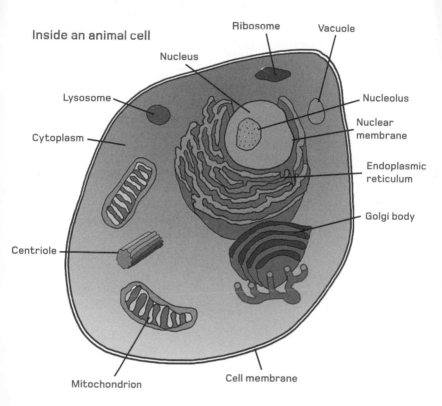

Inside an animal cell

Ribosome

Vacuole

Nucleus

Lysosome

Nucleolus

Cytoplasm

Nuclear membrane

Endoplasmic reticulum

Golgi body

Centriole

Mitochondrion

Cell membrane

Linking genes to DNA

It is impossible to mention DNA without also discussing genes, and vice versa. However, the discoveries of DNA and genes were separate events that were not linked for some time. During the early decades of the 20th century, it was clear that genes carried the information that was passed down through the generations, but the medium by which genes were transferred was unclear. Scientists generally thought that protein was responsible.

In a series of experiments involving bacteria, Oswald T. Avery entertained the possibility that DNA — known since the 1860s — might have a role in transferring information. He discovered that bacteria still transformed into a different strain when proteins were removed. This proved that protein was not transmitting the genes. In a different experiment, DNA was removed, and this time the bacteria did not undergo change. DNA was thus crucial to genetic transfer: genes were made of DNA. This result, published by Avery and his colleagues in 1944, inspired some, but was strangely ignored by others.

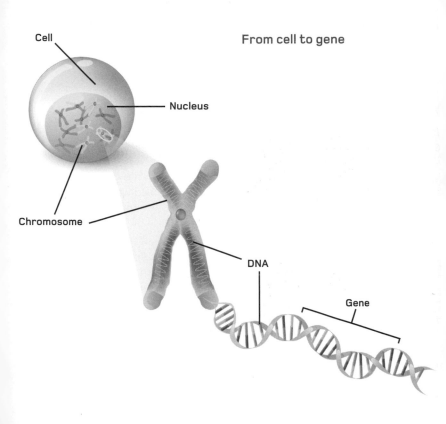

From cell to gene

Cell

Nucleus

Chromosome

DNA

Gene

The discovery
of chromosomes

Chromosomes are packages, located within the nuclei of cells, that contain an organism's complete DNA sequence. Each chromosome is formed of 'chromatins', themselves consisting of DNA strands wrapped around proteins. During the sort of cell division involved in reproduction, chromosomes split apart. These split sections then recombine with the split sections of other chromosomes, a process termed 'genetic recombination'.

Relatively large, chromosomes are visible using ordinary microscopes. Consequently, they were discovered long before DNA and have been known since 1842 when Swiss botanist Karl Wilhelm von Nägeli documented distinct structures within the cell nucleus. These were named by Belgian biologist Edouard Van Beneden in 1888. By about 1900, scientists understood that chromosomes were involved in the transmission of hereditary information and made reference to a 'chromosome theory of inheritance'. Human chromosomes had been counted by the 1920s, though there were some initial errors not corrected until the 1950s.

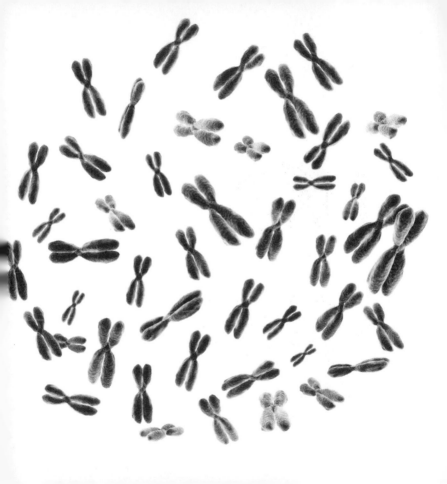

Watson, Crick and Rosalind Franklin

One of the greatest discoveries in genetics was the proposal made by James Watson and Francis Crick in 1953 that DNA was arranged in a double helix, its two strands connected by hydrogen bonds. Their work occurred at the Cavendish Laboratory in Cambridge where they used X-ray data generated by two researchers based in London – Maurice Wilkins and Rosalind Franklin. Their conclusion was revelatory and Watson, Crick and Wilkins received the Nobel Prize in Medicine in 1962.

Watson and Crick (opposite) were not the only scientists working on the structure of DNA at the time. In fact, several researchers before them had established the chemistry of DNA and at least some aspects of its structure. It should also be noted that Watson and Crick did not discover DNA: that credit goes to Friedrich Miescher (see page 72). Rosalind Franklin was key in providing the data used by Watson and Crick and it has been argued that she never received the credit she deserved. Her death in 1958 meant that she did not share the Nobel Prize.

The DNA puzzle

James Watson and Francis Crick are most frequently associated with the realization that the molecule DNA is crucial to the transfer of genetic information. The idea did not result from a single discovery, however, but was the culmination of many studies that had been made beforehand.

Following Miescher's work (see page 72), biochemist Phoebus Levene argued, in 1919, that DNA was formed of numerous nucleotides. His theory, the 'polynucleotide model', was essentially correct, but too simple, as later work would show. A key discovery – made by Erwin Chargaff – was that the sequence of nucleotides varies substantially among species (Leverne thought they were always arranged in the same sequence).

What Watson and Crick did was to use all of this information to establish DNA's three-dimensional structure, combining it with new techniques for working out how molecules could be constructed and modelled.

The DNA molecule consists of nucleotides, each of which involves three components: a sugar, a phosphate and a base or nucleobase (arranged here going from left to right). It took decades to work out how these components were arranged three-dimensionally.

Heterozygosity vs homozygosity

A given gene – that piece of the human genome (the complete set of chromosomes) that carries information about a specific trait – can come in more than one form, and these variants are called 'alleles'. If an individual possesses two different alleles of a given gene, it is said to be 'heterozygous' for that gene. Meanwhile, an individual possessing two identical alleles of a given gene is said to be 'homozygous' for that gene.

Among heterozygous organisms, some alleles are dominant – meaning that they result in the expression of a given trait *whenever* they are present. Other alleles are recessive, meaning that they will only result in the expression of a given trait when a dominant allele is absent. Among homozygous organisms, an individual can possess two dominant versions of a trait (in which case it is termed 'homozygous-dominant') or two recessive versions (in which case it is 'homozygous-recessive'). Typical expressions of heterozygosity and homozygosity concern eye colour in humans and coat colour in domestic mammals.

Labrador coat colour is affected by two genetic traits, known as the B and E loci. Golden labradors always have two recessive 'e' alleles, while chocolate labradors have two recessive 'b' alleles and at least one dominant 'E' allele.

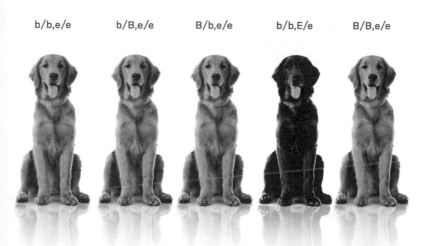

b/b,e/e   b/B,e/e   B/b,e/e   b/b,E/e   B/B,e/e

Genetic drift

Evolution can be a game of chance. Through accident and luck, some organisms will produce more viable offspring than others. A natural disaster or predation event will happen to remove some individuals from a gene pool, and adaptation and fitness will have nothing to do with it. Events of this sort have the capacity to change the evolutionary history of a population and to reduce its original genetic variation.

The phenomenon whereby luck and chance alters the outcome of evolution is known as 'genetic drift'. Sometimes, a substantial amount of original variation is lost in the process and survivors all descend from a subset of the original population (as with the northern elephant seal, whose population was reduced to just 10 around 1900). Such events create what are termed 'population bottlenecks'. Several different mathematical models have been devised to explain and depict genetic drift. These show that random events act faster and have a longer-lasting impact the smaller the population.

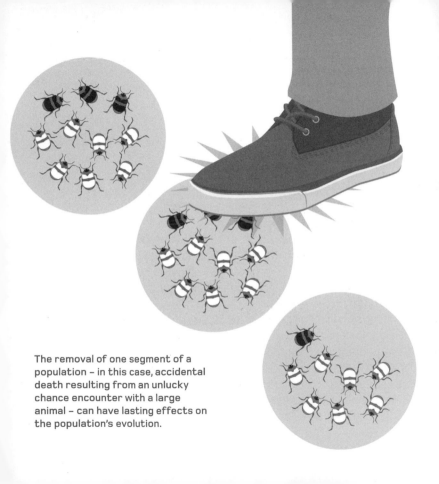

The removal of one segment of a population – in this case, accidental death resulting from an unlucky chance encounter with a large animal – can have lasting effects on the population's evolution.

Homeobox genes

One of the most exciting discoveries in genetics is that of a number of 'master control' genes that determine how the fundamental form of an organism's structure is arranged. These are termed 'homeobox genes'. Generally arranged in clusters, they bind to the rest of the DNA and, via a specific section called the 'homeodomain', activate a cascade of genetic events during early development. Key events activated by homeobox genes include the essential layout of the body and the relative position of limb buds. They also control the growth of cells and thus suppress the formation of tumours.

Homeobox genes were first discovered in fruit flies in 1983. While they are typically discussed within the context of animal development, they also occur in plants, fungi and other groups of organisms. Humans possess approximately 235 genes that fulfil this role, in addition to 65 pseudogenes (genes that are similar in essential structure but do not contain the information to allow the formation of proteins).

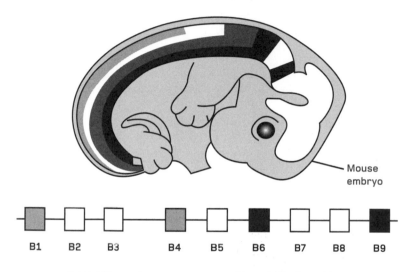

B1　B2　B3　B4　B5　B6　B7　B8　B9

A set of homeobox genes, arranged in a particular pattern, control the sequence in which parts of an organism's structure develop. This colour-coded diagram links specific genes to the areas where they drive development.

Hox genes

Hox genes are a specific set of homeobox genes that control the development and identity of structures along an organism's body axis. Specifically, they activate the development of structures that give a specific body section the identity that it has. They are unique to animals and are best known for the role they play in the development of limbs. Hox genes are arranged in a sequential order and disruption of their order will cause development of the 'wrong' structure in the wrong place: people have bred fruit flies in laboratories, for example, that have legs (instead of antennae) growing from their heads.

Hox genes were first discovered in *Xenopus laevis* (African clawed frogs) in 1984. Their widespread nature in animals indicates that they have existed for virtually the whole of animal history. They are essentially like all homeobox genes in that their basic structure and behaviour is similar across all organisms that possess them. Insertion of Hox genes from, say, a vertebrate into a fly will still permit normal development in the fly.

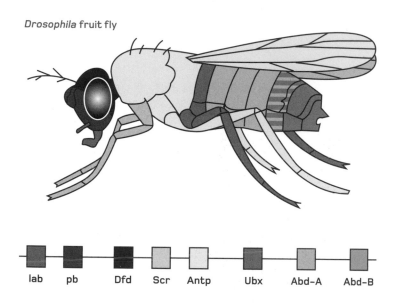

Drosophila fruit fly

lab pb Dfd Scr Antp Ubx Abd-A Abd-B

Hox genes control the positions of structures that grow along the body's axis and are associated in particular with the location of limbs. Colour coding links parts of a fruit fly to specific genes that control development.

Frame shifts

Structures that look similar in related organisms are generally understood to be 'homologous'. That is, they are formed of the same tissues and developed in the same way. Studies of some animals have shown, however, that things are not always this simple. On occasion, a location on an animal's body will grow a structure that, in the ancestor, grew elsewhere, since the genes controlling the expression of that structure have shifted. These are termed 'frame shifts' and are mostly associated with the development of digits. A frame shift means, for example, that what looks like a thumb can grow in the position expected for a second finger. Frame shifts also appear to influence the position of certain vertebrae and teeth in mammals, and have become an important area of debate in discussions of vertebrate evolution.

More broadly in genetics, the term 'frame shift' is associated with a kind of genetic mutation where segments of the genetic code have not been inserted or deleted as they should have, the result being unusual (and often harmful) expression of proteins.

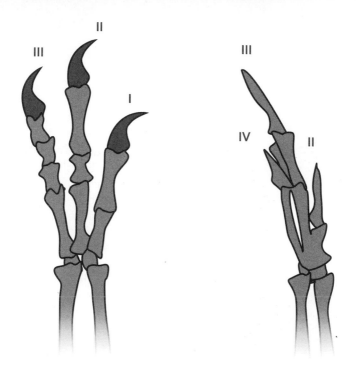

Some experts argue that a frame shift occurred in the dinosaur hand, the location once occupied by digit I suddenly being occupied by a digit that was previously digit II.

Horizontal gene transfer

Evolution mostly concerns the transmission of genetic information down the generations – that is, vertical gene transfer. In recent decades it has been well documented that genes are also – in cases – transferred directly from one organism to another in events that do not involve reproduction. Instead, one organism might be injected with the genes from another, typically via the actions of a parasite. The new genetic information is then assimilated, altering the genetic code. This phenomenon is called 'horizontal gene transfer' (HGT). It was first documented in 1951 in the bacterium *Corynebacterium diphtheriae*, the species that causes diphtheria.

Widespread among bacteria, HGT involves species that belong to separate branches on the family tree. There are also indications that it has occurred among more complex organisms, including insects and mammals. The emerging view that HGT is frequent and widespread has led to suggestions that we should be talking about a 'web of life' rather than a tree (see page 102).

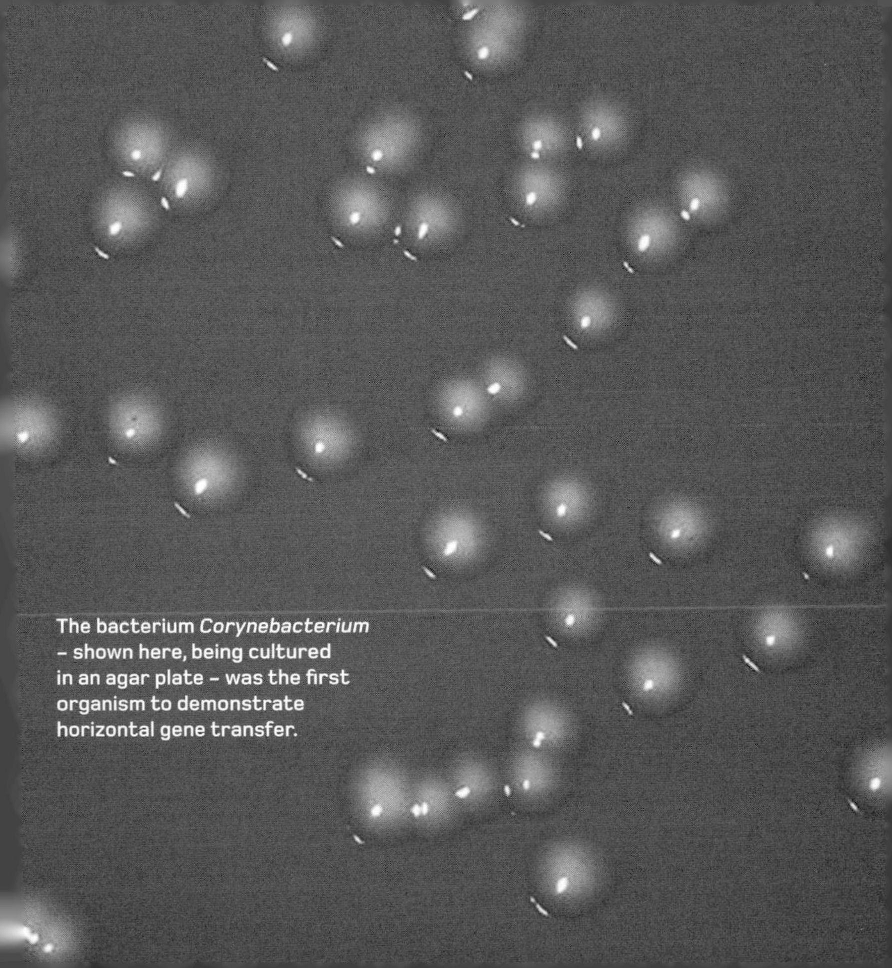

The bacterium *Corynebacterium* – shown here, being cultured in an agar plate – was the first organism to demonstrate horizontal gene transfer.

The selfish gene

The discovery that information is passed down the generations via genes has resulted in the view that all 'decisions' involved in biology – conscious and unconscious – are done to benefit genetic fitness and the successful transmission of genes. According to this view, one organism might help others because they share genes and are assisting in the perpetuation of their own genetic group: altruism is never really a selfless act, therefore, but one in which an organism assists its genetic group somehow. This idea is termed 'kin selection theory'. The idea that such decisions made by organisms do, in fact, reflect genetic 'selfishness' was famously discussed in the 1976 book *The Selfish Gene* by Oxford-based scientist and author Richard Dawkins.

The 'selfish gene' is widespread in biology, but it might not be the only explanation for genetic persistence. Biologists have long noted that groups work together because complex relationships require the perpetuation of numerous individual components, not all of which will persist if kin selection alone is in operation.

Evo-devo

A combined understanding of the fossil history of modern organisms, the development of their embryos, and the genetics that control their development has resulted in a field known as 'evolutionary development biology', or 'evo-devo' for short. It basically combines everything we know about the evolutionary past of an organism in order to understand its development as a living organism. It shows how the pattern of evolution has been shaped by the process of development. If we understand that certain body parts, for example, appear in a given developmental sequence, we gain insight into evolutionary history.

The term 'evo-devo' is certainly modern; as a distinct discipline, the field itself is generally regarded as having arisen since the 1990s. The 1980s discovery of homeobox genes (see page 210) was among the key events that helped establish evo-devo as a distinct field of research, as was the sequencing of entire genomes. The field as a whole, however, has roots that go all the way back to the 19th century.

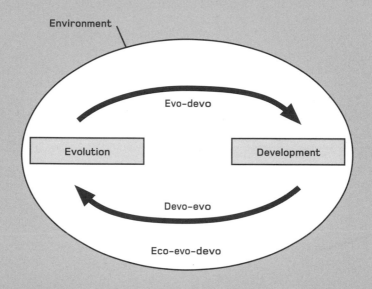

This schematic depicts links
between evolution, development
and environment as studied in
evo-devo research.

Gene sequencing

Once the basic structure of DNA became known, people began to determine its sequence of genetic and chemical components. The specific arrangement of chemical components varies from one organism to another, and from one individual to another. Consequently, DNA collected from organisms, or from organic information they leave behind in the environment, can be used to identify a species, and even an individual of that species. This last fact is key to modern forensic work, in tracking missing people, and in identifying parentage.

Any technique that involves determining the sequences of nucleotides in DNA is termed 'gene sequencing'. It can involve segments of DNA or even the entire genome. Sometimes a particular DNA sequence is duplicated so that other copies are obtained for additional study. Numerous different machines – DNA sequencers – are now available for sequencing. The first of them was invented in 1987. As technology and techniques have improved, sequencing has become faster and cheaper.

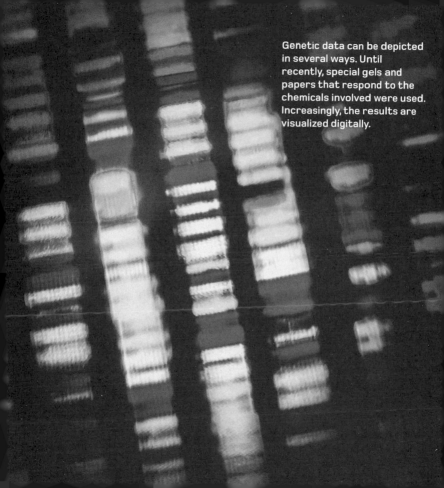

Genetic data can be depicted in several ways. Until recently, special gels and papers that respond to the chemicals involved were used. Increasingly, the results are visualized digitally.

The human genome

A full understanding of human genetics is key to several lines of enquiry. It allows us to discover which parts of our genetic code are linked to disease and other health issues; to discern what our genome says about our evolutionary past; and to learn how our genes are linked to our development, anatomy, biology and behaviour. Consequently, considerable effort has been devoted to mapping the entire human genome.

In 1990, a team of scientists embarked on the Human Genome Project (HGP), an enormous collaborative involving teams around the world. The project took 13 years and cost $3 billion. Initial results were announced in 2000 and completion was declared by 2003 (though not really achieved until 2006). Thanks to the HGP, we now know that there are around 20,500 genes in our species, and we know where on the genome they are located and what their functions are. Numerous other organisms have had their entire genomes sequenced since 2000. They include several insects, a pufferfish, rice and various microorganisms.

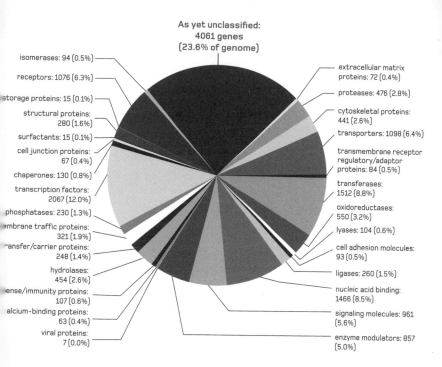

As yet unclassified:
4061 genes
(23.6% of genome)

isomerases: 94 (0.5%)

receptors: 1076 (6.3%)

storage proteins: 15 (0.1%)

structural proteins:
280 (1.6%)

surfactants: 15 (0.1%)

cell junction proteins:
67 (0.4%)

chaperones: 130 (0.8%)

transcription factors:
2067 (12.0%)

phosphatases: 230 (1.3%)

membrane traffic proteins:
321 (1.9%)

transfer/carrier proteins:
248 (1.4%)

hydrolases:
454 (2.6%)

defense/immunity proteins:
107 (0.6%)

calcium-binding proteins:
63 (0.4%)

viral proteins:
7 (0.0%)

extracellular matrix
proteins: 72 (0.4%)

proteases: 476 (2.8%)

cytoskeletal proteins:
441 (2.6%)

transporters: 1098 (6.4%)

transmembrane receptor
regulatory/adaptor
proteins: 84 (0.5%)

transferases:
1512 (8.8%)

oxidoreductases:
550 (3.2%)

lyases: 104 (0.6%)

cell adhesion molecules:
93 (0.5%)

ligases: 260 (1.5%)

nucleic acid binding:
1466 (8.5%)

signaling molecules: 961
(5.6%)

enzyme modulators: 857
(5.0%)

This diagram shows the numerous components of the human
genome, categorized by function. The numbers show how many
genes are involved and what percentage of the genome they occupy.
As is clear, much work remains to be done.

The origin of life

How life originated remains one of the greatest questions in biological research. The specific event that led to the origin of life – the event termed 'abiogenesis' – cannot have involved the development of one organism into another, and thus was not an evolutionary event. The origin of life is not, therefore, a question that evolutionary biologists can answer. It might be that the earliest organisms were microscopic single cells or 'protocells' consisting of an external membrane, an internal jelly, a power source and a molecule that could be duplicated. The idea that such structures originated by chance seems absurd. Several experiments have shown how such structures might have arisen, since mineral and organic materials will form cell-like structures given the right conditions. It is currently thought that self-replicating molecules existed before anything like a cell did, however, and that these molecules – perhaps formed of RNA (see page 228) – replicated by attaching to freely occurring chemical bases to form additional molecules. Occasional errors in the duplication process would have started evolution.

Several different environments have been proposed as places where life on Earth might have originated. They include deep-sea vents, hot springs and tidal pools.

The RNA world hypothesis

Discussions of the origin of life have generally assumed that the self-replicating molecule DNA was present from life's origin and that it has always served as the way in which information was stored and transmitted. More recently, however, it has been argued that another, related molecule – RNA (ribonucleic acid) – took this role early in the history of life. The key evidence for this idea comes from the fact that the core parts of various cells are RNA-dominated. Furthermore, manipulation in the laboratory has shown how RNA strings will pair with free nucleotides, form duplicates of themselves, and evolve.

The idea that early life was RNA-themed, not DNA-themed, is known as the 'RNA world hypothesis'. The idea that RNA might have been an antecedent to DNA is itself not new, having originated in the 1960s, but the term 'RNA world' was not used until 1986. DNA appears superior to RNA as far as information storage and transmission goes, since it is more stable, and this presumably explains why DNA became increasingly important.

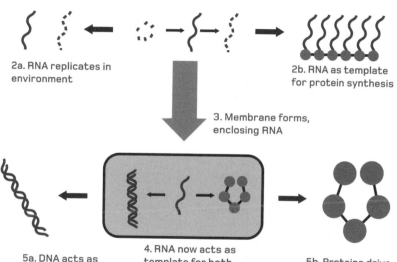

1. RNA forms from inorganic chemicals

2a. RNA replicates in environment

2b. RNA as template for protein synthesis

3. Membrane forms, enclosing RNA

5a. DNA acts as improved master template

4. RNA now acts as template for both protein and DNA synthesis

5b. Proteins drive cellular activity

The Archaean and
the oldest fossils

Geological evidence shows that life was present on Earth by around four billion years ago (bya), during the Archaean. The evidence used to demonstrate the presence of living things at this time does not come from fossils directly, but from the chemistry of ancient rocks. Evidence for oxygen in the atmosphere of 3.7 bya is indicative of living things, since this is the only way oxygen could have been introduced into the Archaean atmosphere. More controversial chemical traces suggest an even earlier presence of life, at around 4.1 bya. The oldest true fossils – structures, termed 'stromatolites', formed of layers of preserved microorganisms – date to 3.5 bya. These discoveries show that life originated early in the history of our planet (Earth itself originated 4.5 bya). This is consistent with the fact that numerous modern microorganisms are specialized for life in extreme environments – those that are too toxic or too hot compared to those inhabited by the majority of multicellular organisms. It appears that such environments were typical of the oldest organisms and even favoured by them.

Stromatolites are peculiar structures consisting of sheet-like layers of microorganisms. They are among the oldest indications of life on Earth.

1 cm

The symbiotic cell

Eukaryotic cells – those common to all multicellular forms of life – are made complex relative to those of prokaryotes in that they contain several internal structures, called 'organelles'. Because these organelles resemble various prokaryotes that live independently, it has long been thought that the very first eukaryote arose after several free-living prokaryotes combined and began living as a united organism. Such acts – known as 'symbiosis' – are common across nature and have evolved on hundreds of separate occasions. The most popular symbiotic model involved in eukaryote origins is termed the 'serial endosymbiotic theory' and was proposed by Lynn Margulis in 1967.

The advantage of symbiosis for the organisms concerned is that their different abilities allowed them, collectively, to deal with the toxic chemical conditions present in the environments of the early Earth. Mitochondria appear to be derived from oxygen-using bacteria that would have allowed the symbiotic cell to survive in oxygen-rich environments, for example.

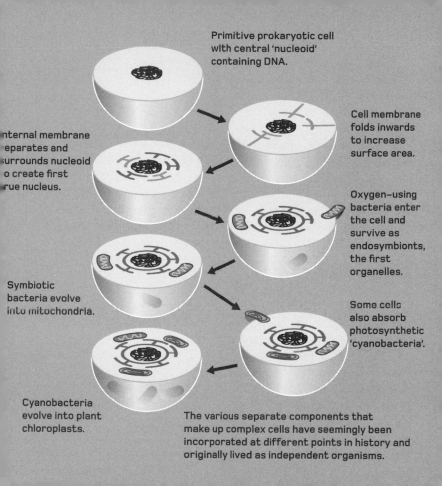

Primitive prokaryotic cell with central 'nucleoid' containing DNA.

Cell membrane folds inwards to increase surface area.

Internal membrane separates and surrounds nucleoid to create first true nucleus.

Oxygen–using bacteria enter the cell and survive as endosymbionts, the first organelles.

Symbiotic bacteria evolve into mitochondria.

Some cells also absorb photosynthetic 'cyanobacteria'.

Cyanobacteria evolve into plant chloroplasts.

The various separate components that make up complex cells have seemingly been incorporated at different points in history and originally lived as independent organisms.

The rise of sponges

Sponges are peculiar aquatic animals that live rooted to a particular spot, a mode of lifestyle known as 'sessility'. They essentially consist of a porous mass of tissue, though a supporting skeleton of protein fibres or silica shards is usually present. Digestive and circulatory systems are absent and sponges rely on water circulating through their numerous tiny pores and channels. They feed mostly on bacteria. The majority of sponges are marine but there are freshwater species too.

Sponges are one of the oldest animal groups – genetics show that they probably diverged prior to 600 million years ago (mya). Numerous fossils are known from Cambrian rocks, and sponge reefs, possibly similar in size to modern coral reefs, were present in the Jurassic. More mobile animals originated around the same time, including cnidarians and comb jellies, but most experts think sponges diverged first from the common animal ancestor, in which case their simple anatomy and lifestyle likely represent 'primitive' features for the entire Kingdom Animalia.

Fossils of Ediacara

Ancient rocks worldwide – dated between 610 and 540 million years old – yield weird, flattened, soft-bodied fossil animals, most of which seem to have been bottom-dwellers. They are disc-like, tubular or have quilt-like structures, and range in length from a few centimetres to 1 m (40 in). Named after the Ediacara Hills of South Australia, these are the ediacarans.

Ediacarans were in existence some time after complex, multicelled organisms originated 800 mya, but before the major modern animal groups that include insects and vertebrates and their kin originated – 540 mya. A distinct stage of Earth history, the Ediacaran Age, takes its name from this part of geological history. Classifying ediacarans has proved controversial. Some resemble groups present today, such as worms, and some experts propose that ediacarans include early members of these living groups. Others argue that ediacarans are unlike living organisms and more likely represent an early branch of animal history that died out without descendants.

Dickinsonia – an oval fossil with numerous segments – is a well known ediacaran. Specimens range from just a few millimetres in length to over 1 m (40 in).

Cnidarians and comb jellies

Several groups of soft-bodied animals occur worldwide in marine and freshwater environments, sometimes in phenomenal abundance. Among these are cnidarians – a group that includes more than 10,000 species of corals, sea jellies, sea anemones and hydras – and ctenophores, or comb jellies. Some animals often regarded as sea jellies or jellyfish are not single organisms, but colonies – for example, the Portugese man o'war, a member of the siphonophore group. Comb jellies are similar to cnidarians and might be their close relatives. They are typically rounded or egg-shaped marine predators that swim using rows of tiny hairs, termed 'cilia'.

The oldest cnidarian fossils date from the Ediacaran, around 600 mya, and from rocks older than those bearing ediacaran fossils themselves. Both anatomy and DNA show that cnidarians and comb jellies originated very early in the history of animals, and were among the first groups to have diverged from the earliest, ancestral animals. They seem to have originated about 700 mya.

The Burgess Shale and the Cambrian Explosion

One of the world's most famous fossil-bearing rocks is the Burgess Shale in British Columbia, Canada. It was laid down during the Cambrian, about 510 mya, when mud slides buried bottom-dwelling marine animals. The fossils are often complete and with their soft parts preserved. They show that – by the middle part of the Cambrian – animals had diversified into all of the major groups. This diversification event began about 540 mya and is known as the 'Cambrian Explosion'.

Trilobites (see page 248) are present in the Burgess Shale, as are other bottom-feeding arthropods and worms. Animals with flap-like fins or fleshy legs are early members of the arthropod lineage. They include the spiny *Hallucigenia* and *Opabinia*, a five-eyed animal with a proboscis. Most animals were less than 15 cm (6 in) long, but *Anomalocaris* – a swimming predator with a circular mouth and arm-like feeding appendages – reached 1 m (40 in). Similar fossil assemblages in China and Greenland show that such animals were present worldwide by 520 mya.

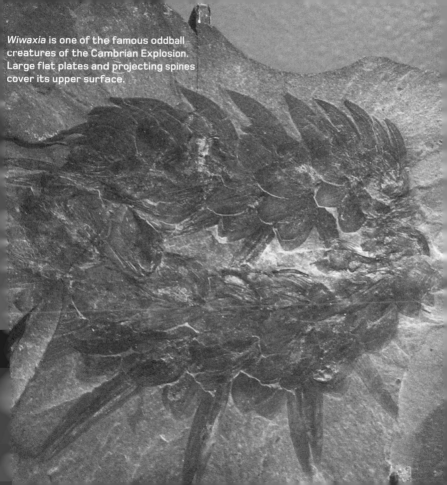

Wiwaxia is one of the famous oddball creatures of the Cambrian Explosion. Large flat plates and projecting spines cover its upper surface.

Shells and shelly fossils

Hard, mineralized shells that cover and protect the soft bodies of invertebrate animals are a typical feature of the modern world. Shelled invertebrates are important filter-feeders, predators and grazers on the sea floor as well as in rivers, lakes and terrestrial environments. Armour plates, spines and other structures protect the bodies of numerous other animals too.

Fossils show that shelled invertebrates originated between 550 and 530 mya, during the Ediacaran and Early Cambrian. The appearance of animals with such hard parts was a key event. Various tiny cone- or tube-like fossil shells are known from this time and form what are known as 'small shelly faunas'. The fossils have shell lengths of just a few millimetres. The exact appearance of these animals is uncertain, but it is assumed that they were similar to modern shelled molluscs. Other shelly fossils from the period include tiny armour plates, hundreds of which covered a variety of animals including a peculiar group of early armoured slug-like animals called the halkieriids.

The so-called 'small shelly fossils' of the Ediacaran and Cambrian are mostly fragmentary and sometimes difficult to interpret. They have been discovered worldwide.

Plants invade the land

Land-living plants are essential to terrestrial ecosystems today, and have been for more than 350 million years. Prior to this time, plants were restricted to aquatic environments. Plants did not invade the land just once, but different groups invaded the land at different times. The result was a 'staged' invasion that occurred between 500 and 300 mya.

The first plants to move onto land were algae, the group that includes microscopic, planktonic forms as well as seaweeds. Species formed films, or mats, at the edges of pools and streams, and later across mudflats and shorelines. By about 450 mya, during the Ordovician, fossil spores show that plants similar to modern liverworts were living on land. Like algae, these would have been dependent on constant access to water and would have dried out if a water covering was not available. Truly terrestrial plants appeared around 430 mya, during the Silurian. The oldest is *Cooksonia*, a tiny, simple plant just a few millimetres high.

Cooksonia lacked leaves and roots and had a simple branching form. Cap-like structures at the tips of its branches released spores.

Sea scorpions and other arthropods

One of the largest and most important animal groups is Arthropoda, moulting, jointed-limbed animals that include arachnids, crustaceans and insects. Millions of species alive today evolved from marine ancestors that originated over 550 mya. All modern arthropod groups had evolved by the end of the Devonian (about 360 mya).

Among the most spectacular of fossil arthropods are the eurypterids or sea scorpions, a group that existed between the Ordovician about 470 mya and the Permian about 250 mya. In general, they had large, flattened anterior sections; a long, tapering tail region; enlarged, flattened limbs used in swimming; and anteriorly positioned walking and grabbing limbs. The largest species exceeded 2 m (6½ ft) in length, but most were less than 30 cm (12 in) long. Large, powerful-looking pincers suggest that some eurypterids were dangerous predators. However, they most likely consumed small arthropods and may also have been scavengers. Some species were capable of walking on land.

Key eurypterid features include paddle-like limbs, walking legs covered in hairs, a long spike at the tail tip and a broad, flattened body.

The age of trilobites

Sea-dwelling invertebrates, trilobites existed for more than 250 million years, disappearing in the extinction event that ended the Permian (see page 286). This group's oldest members lived during the Early Cambrian, about 520 mya. More than 5,000 distinct kinds have been named. A head shield (the cephalon) is present, then a thorax consisting of numerous flexible segments, and a tail segment (the pygidium). The body consists of three longitudinal sections: a (typically) convex midline section with two down-curved sections either side; this three-part arrangement explains the group's name ('trilobite' meaning 'three-lobed').

Trilobites were the oldest arthropods to possess the hard external covering typical for the group. They had antennae and numerous jointed limbs. A number of lifestyles and body shapes evolved within the group. The majority were bottom-dwelling scavengers, predators or filter-feeders; some also swam. Long spines or horn-like structures evolved, and big complex eyes were typical. Some burrowing trilobites lacked eyes entirely.

Echinoderms

Among the most abundant and diverse of marine animals are the echinoderms, a weird group that typically possess radial or pentameral symmetry, this meaning that the body can be split into five equal-sized sections when viewed from above. Other key features include tentacle-like, sucker-tipped tube feet controlled by pressurized water, and amazing regenerative abilities.

Today's species include sea lilies, brittle stars, sea stars, sea cucumbers, sea urchins and sand dollars. Some are super-abundant in certain environments and crucial as prey items and consumers on the sea floor. Many are soft-bodied but others contain hard skeletons formed of interlocking plates. The majority of fossil echinoderms were close kin of these living kinds and fossils show that they were essentially similar to them. A few extinct crinoids were spectacular, however, in particular the giant sea lilies of the Mesozoic. Anatomical, genetic and fossil evidence shows that echinoderms are close relatives of chordates and hemichordates (see page 254).

Sea lilies, or crinoids, possess a 'crown' section where arms emerge from a cup or calyx. Complete fossils like these are similar to modern species.

Graptolites

Many peculiar animals inhabited prehistoric oceans, among them the graptolites, a group that existed between the Cambrian and Carboniferous. Typically just a few centimetres long and rod-shaped, branching or netlike, graptolites are not the skeletons of individual organisms but are colonies of tubes or cup-like structures, each of which was inhabited by a small, soft-bodied animal (termed a 'zooid'). Rare fossils show that zooids had slender arms, each covered in miniscule tentacles, used for grabbing food particles from suspension.

Graptolite fossils are extremely common, the members of individual species are geographically widespread and species evolved quickly and were short-lived (lasting no more than one million years). As a consequence of these factors, graptolites serve as 'index fossils': because a given species tends to be characteristic of one particular geological time zone, the discovery of that species in rocks anywhere in the world allows scientists to know the age of that rock layer.

Chordates and hemichordates

Mammals, reptiles, amphibians and fish are vertebrates, a group with an internal skeleton, components of which are formed of a mineralized tissue we call bone. Several surviving animal groups are close relatives of vertebrates and show what their ancestors would have been like. One group lack bones but do possess a flexible rod that extends along the body (the notochord). A modified version of this structure became part of the spine in vertebrates. All notochord-bearing animals are chordates, and one particular group, the tiny, sediment-dwelling lancelets, are close relatives of animals that, over 500 mya, gave rise to vertebrates.

The fossil record, anatomy and genetics of living animals show that chordates and vertebrates are part of a larger group, termed 'deuterostomes'. Also among deuterostomes are the echinoderms (see page 250) and the poorly known, worm-like hemichordates. Despite overall appearance, hemichordates are unmistakeably chordate-like, possessing a notochord-like structure and slit-like gills that link them to that group.

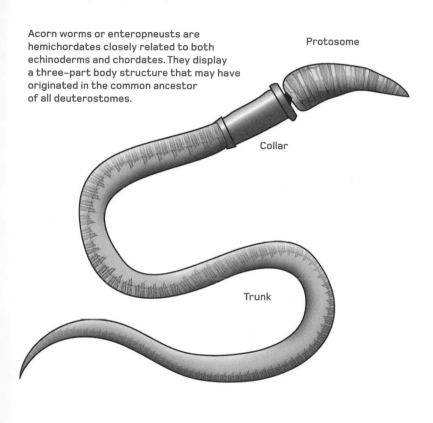

Acorn worms or enteropneusts are hemichordates closely related to both echinoderms and chordates. They display a three-part body structure that may have originated in the common ancestor of all deuterostomes.

Protosome

Collar

Trunk

Arthropods invade the land

When we speak of animals 'invading the land', we normally have four-footed tetrapods in mind (see page 278). But two other major groups of organisms had already made the transition to terrestrial life: plants and arthropods. Jointed-limbed animals with hard external skeletons, arthropods include insects, arachnids, crustaceans, millipedes and centipedes.

Millipedes, centipedes, scorpions, spiders and kin, crustaceans and insects represent independent invasion events from the aquatic environment, each group evolving land-going adaptations separately and at different times between 450 and 400 mya. Late Ordovician trackways indicate that millipedes were the first terrestrial invaders. Land-living mites, centipedes and insects are known from Devonian rocks about 390 million years old, by which time arthropods of many kinds were well established on land. By the end of the Devonian, they were eating rotting plant material, consuming carrion and preying on other arthropods. They were to become the most abundant land-living animals of all.

One of the most remarkable
of ancient arthropods
is the gigantic millipede
Arthropleura. It lived about
300 mya and reached 2.3 m
(7.5 ft) in length.

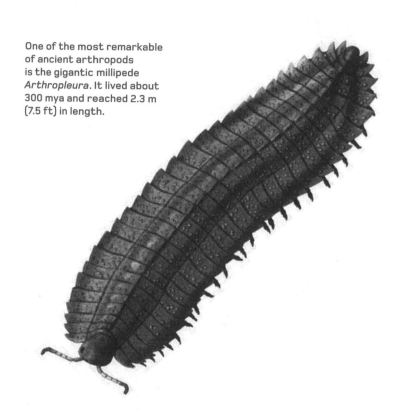

Conodonts and other early vertebrates

Around 515 mya, in the Cambrian, early chordates gave rise to a new group with large, complex eyes and a structure formed of a novel material: a bony skull. These were the first vertebrates, the group that later gave rise to cartilaginous and bony fishes. The earliest vertebrates possessed a single nostril and their mouths were simple openings absent of mobile bony parts. For this reason they are termed 'jawless fishes' or 'agnathans'. One agnathan lineage survives today: the cyclostomes, the group that includes the predatory and parasitic hagfishes and lampreys.

Other jawless fish groups originated at about the same time as cyclostomes: conodonts. These eel-shaped animals ranged from 4–40 cm (2–16 in) and were long known only from their tiny, slender teeth. Cyclostome- or conodont-like animals gave rise to new agnathan groups where the body and tail were shorter and the head was covered with large bony shields. These shields were ancestral to the large skull bones of later vertebrates. Pectoral fins and taller, muscular tail fins also evolved in these animals.

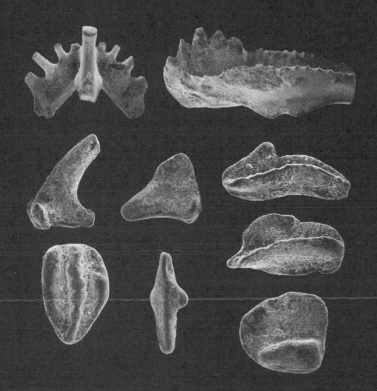

Conodonts are best known for their tiny, tooth-like elements. These are extremely abundant in many ancient rock layers.

The vertebrate skeleton

Vertebrates are those animals that possess a bony internal skeleton comprising a skull, a vertebral column formed of numerous separate vertebrae, and a set of fins or limbs connected to bony girdles. The very earliest vertebrates – small jawless fishes, less than 10 cm (4 in) long – did not have a bony skeleton at all, but one formed of cartilage. Their spinal cord was protected by a row of cartilaginous discs, and fins and their girdles were absent. Over time, hard plates formed in the skin around the head, perhaps for defence or structural strength. These merged to form a protective casing for the whole of the head, the result being a skull.

By about 430 mya, fishes with a complete bony skull, a row of bony vertebrae, and bones in their fins and fin girdles had evolved. The vertebrate skeleton had therefore been compiled, but it had been done so in piecemeal fashion, the final structure being a composite of structures grown both deep within, and on, the outer edges of the body.

Early jawless fishes, such as the streamlined Devonian *Pteraspis* shown here, combined large, shield-like head plates with a cartilaginous internal skeleton.

The evolution of jaws and teeth

By the Silurian (around 440 mya), vertebrates of many kinds swam in the world's oceans, rivers and lakes. All lacked a feature typical of vertebrates today – and integral to the way in which they capture and process food – jaws. Their appearance resulted in the evolution of vertebrates able to consume shellfish or kill other large animals, since they allowed vertebrates to grab, engulf and crush food. It is generally thought that jaws are modified versions of the bar-like bones used to support the gill region. Jawed vertebrates are called 'gnathostomes'.

Shortly after the appearance of jaws, a second anatomical feature evolved: teeth. These are hard structures where external layers of enamel and dentine provide cutting, shearing or chewing surfaces while surrounding an internal pulp cavity. They either originated from tooth-like spines (denticles) originally located across the outside of the body, or from enlarged, enamel-coated scales. Either way, gnathostomes with toothed jaws were widespread by the late Silurian, 400 mya.

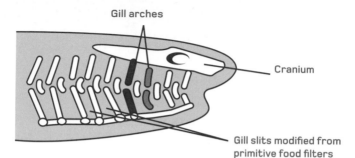

Gill arches

Cranium

Gill slits modified from
primitive food filters

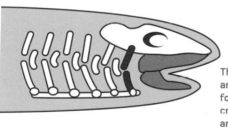

The jaw evolved as bony gill
arches gradually shifted
forward and attached to the
cranium to create an upper
and lower jaw around a newly
formed mouth. Subsequently,
a second set of gill arches was
also incorporated to act as a
strengthening brace.

Shelled swimming molluscs

Molluscs originated in the Cambrian as small, bottom-dwelling animals with a muscular foot and a protective shell. They became extremely diverse, giving rise to twin-shelled bivalves (clams, mussels, oysters), burrowing tusk shells and numerous other subdivisions. Over 80,000 species are alive today.

Perhaps the most remarkable molluscs are the cephalopods, a group that evolved a distinct head region, prominent arms and an ability to swim by pumping water out of a tube-like structure called a 'siphon'. Most ancient cephalopods were protected by a large shell that was either long and bullet-shaped or coiled. Coil-shelled cephalopods – in particular, the ammonites – thrived throughout the Jurassic and Cretaceous, becoming among the most abundant of marine animals. They ranged in size from a few centimetres to over 2.4 m (8 ft) across. While it is generally thought that ammonites became extinct at the close of the Cretaceous, some experts argue that octopuses evolved from ammonites at some point during that same period.

The age of fishes

Around 450 mya, jawless agnathan fishes (see page 320) gave rise to a new group: the jawed fishes or gnathostomes. Jaws likely evolved from strut-like bones originally part of the gill apparatus. They allowed newly evolved fish groups to adopt many lifestyles unexploited by agnathans. So many new jawed fishes were in existence by the Devonian (between about 420 and 360 mya) that some experts dub this time the 'Age of Fishes'.

Among the jawed fishes were various placoderms: those that combined a flexible tail with a front half either heavily encased in boxlike armour; raylike bottom-dwellers with pectoral fins that resembled arthropod limbs; and large predators that inhabited Devonian seas. Placoderms probably included the ancestors of all later jawed fishes, also abundant during the Devonian. Chondrichthyans, or cartilaginous fishes, had given rise to shark-like forms, and others, by this time, some of which had giant slicing teeth or resembled modern rays and frogfishes. Many of these animals died out in an extinction event that ended the Devonian.

The giant Devonian placoderm *Dunkleosteus* is one of the most famous of fossil fishes. It was clearly a top predator in the Devonian seas and reached 6 m (20 ft) in length.

Spiny jawed fishes

One of several important fish groups of the Devonian, the acanthodians were best known for the spines that grew from their bellies and fins. For years these animals were known as 'spiny sharks'. It is more typical today to refer to them as 'spiny jawed fishes' – jawed fishes with spines, not fishes with spiny jaws. Acanthodians evolved during the Ordovician, probably from placoderms (see page 320), and died out during the Permian. They ranged from 10 cm–2.5 m (4 in–8 ft) in length; most were streamlined but some were deep-bodied and others almost eel-like. While some possessed as many as ten sets of spines, others lacked spines entirely. The majority either fed on small food particles in the water or on invertebrates or smaller fishes.

Recent discoveries have shown that the pectoral and anal fin spines, long thought unique to acanthodians, were inherited by the earliest cartilaginous fishes. Indeed, the very oldest cartilaginous fishes we know of were highly similar to acanthodians overall and seemingly evolved from this group approximately 440 mya.

Acanthodians like this *Nerepisacanthus* were streamlined predators of both marine and freshwater environments. They were an important early group of jawed vertebrates.

Fleshy-finned fishes

Fleshy-finned or lobe-finned fishes, properly called sarcopterygians, are relevant to the history of the tetrapods. Sarcopterygian pectoral fins have a muscular base and a robust skeletal framework. These structures are not necessarily used for underwater walking, or anything like it but are ancestral to features inherited by tetrapods. Functional lungs are also present in some living sarcopterygians and were presumably widespread in extinct ones.

Two sarcopterygian groups are present today: the tropical, mostly freshwater lungfishes of South America, Africa and Australia, and the deep-sea coelacanths. Coelacanths owe their fame to the 1936 discovery of the modern *Latimeria* in the western Indian Ocean, and prior to this were mostly associated with the Mesozoic. Around 360 mya, predatory, shallow-water sarcopterygians evolved a long, flattened head where the eyes were placed well back, and more muscular, thicker-boned pectoral fins. These were the ancestors of tetrapods.

Modern coelacanths, such as the preserved specimen shown here, are large, deep-sea fish with blue or bronze scales. Extinct coelacanths were diverse in size, shape and lifestyle.

Insect wings and insect flight

Fossils show that early winged insects, termed 'pterygotes', had evolved from wingless ancestors by the Early Carboniferous, about 350 mya. Wings might have evolved as novel structures – as flap-like extensions on the back, for example – or might be modified gill flaps originally attached to the legs. The genes involved in wing development are similar to those associated with the development of gills, so the latter seems likely.

The context in which wings evolved remains debated. The earliest pterygotes lived in aquatic environments, so one idea is that wings originated as small flaps that helped these animals skim across the water's surface. Another idea is that wings helped early insects glide among tall vegetation. Either way, insects with two or three pairs of non-folding wings lived during the Carboniferous, including early relatives of mayflies, cockroaches and dragonflies. As they adapted to new lifestyles, later insect groups evolved complex, folding wings, but lost some to retain a single pair only or became entirely wingless.

Griffinflies

The term 'griffinfly' applies to the Meganisoptera group of the Carboniferous and Permian. These insects were predatory and had two pairs of enormous, rigid wings with wingspans of 70 cm (28 in). The body was long and slender and the large mouthparts had serrated inner edges. The best known griffinfly is *Meganeura* from the Carboniferous of France. Griffinflies are part of a larger group, the only surviving members of which are dragonflies and damselflies.

What allowed such large insects to evolve in the Carboniferous and Permian? One idea is that the high level of oxygen present at the time encouraged their evolution, another that the lack of vertebrate predators allowed them to evolve. Some suggest that these animals became huge in order to protect themselves from the oxygen poisoning that threatens the larvae of insects that live in oxygen-rich water. Large larvae are less liable to oxygen poisoning than small ones because their relatively small surface area reduces the amount of oxygen their bodies absorb.

Meganeura is known from excellent fossils. It would have looked dragonfly-like overall, with more obvious jaws, a longer head and super-sized wings.

The first trees

Plants invaded the land during the Ordovician (see page 244). By the Late Devonian (approximately 380 mya), many land plants existed. The evolution of waterproofed outer layers, bark, openings in the leaves that allowed the exchange of gases, and internal tubes enabling the transportation of water to tissues, meant that plants could increase substantially in size.

By the Late Devonian there were abundant large trees, some of which were over 30 m (98 ft) tall and 2½ m (8 ft) wide at the base. The term 'tree' does not correspond to a specific group of plants, but to any large land plant with woody bark and branches located some distance from the ground. Dense forests formed for the first time, the rotting leaves and trunks forming thick soil layers that eventually transformed into coal. Coal formation began in the Devonian but is mostly associated with the following Carboniferous age. The most famous Devonian tree is *Archaeopteris*, an early member of the conifer lineage. Other Devonian trees were members of the fern and club moss lineages.

The giant tree *Archaeopteris* was conifer-like overall, but possessed fern-like leafy fronds.

Tetrapods invade the land

Robust pectoral fins, air-breathing abilities and shallow-water, predatory lifestyles allowed some sarcopterygian fishes to survive in shallow, weed-choked habitats close to the water's edge. During the Devonian, some of these animals took to hunting on land, eventually giving rise to tetrapods – the four-footed vertebrate group that includes amphibians, reptiles, birds and mammals. Several Devonian fossils are near-perfect evolutionary intermediates between ancestral sarcopterygians and early tetrapods. The most famous of these 'fishapods' is *Tiktaalik* from Ellesmere Island, Canada, named in 2006.

Trackways show that tetrapods were walking on coastal muds during the Middle Devonian, about 395 mya, and good tetrapod skeletons are known from rocks 370 million years old. The most famous are Greenland's *Acanthostega* and *Ichthyostega* – the latter a large predator (about 1 m/40 in long) that may have moved on land by using its strong forelimbs in crutch-like fashion. These animals had as many as eight fingers and seven toes.

The earliest tetrapods, like *Ichthyostega* shown here, were mostly aquatic. Their strong skulls and large teeth show that they were predators, probably of fish.

The temnospondyl empire

For more than 220 million years, a large group of tetrapods inhabited freshwater and coastal ecosystems worldwide: the temnospondyls. Although typically regarded as amphibians, they were mostly very different from living amphibians and usually much larger. Some species were 10 cm (4 in) long, whereas others exceeded 2 m (6½ ft) and, in some cases, reached 7 m (23 ft).

The majority of temnospondyls resembled large, long-snouted salamanders or broad-headed crocodiles. Aquatic and amphibious lifestyles were common – temnospondyls were among the most important groups of aquatic predators before crocodile-line archosaurs evolved. Long, slender, multi-toothed jaws evolved several times during their history. Members of some groups had very short snouts and comically wide heads. Important throughout the Carboniferous, Permian and Triassic, temnospondyls dwindled during the Jurassic and Cretaceous. However, members of one group – the dissorophoids – very likely include the ancestors of living amphibians.

Temnospondyls, such as this *Eryops*, were large, often formidable aquatic predators with a powerful bite. Thick bony ridges perhaps helped conduct stresses.

The rise of amniotes

During the Devonian and Carboniferous, numerous tetrapods evolved from *Ichthyostega*-like ancestors. The majority were amphibious or aquatic and are generally termed 'amphibians' even though they were mostly very different from the living animals that go by that name. Several of these groups became adapted for life on land, and approximately 315 mya evolved eggs where a membrane (the amnion) helped protect the embryo from the outside world. An eggshell (originally probably flexible and leathery) evolved at about the same time. Eggs of this sort are termed 'amniotic eggs' and their evolution was likely key to the success of the tetrapod group that possessed them: the amniotes.

Amniotes diverged into two lineages: synapsids (the group that includes mammals) and reptiles. Fossils show that the early members of both groups looked similar. Shelled eggs were lost many times during amniote history: numerous reptiles give birth to live babies, and shelled eggs are rare in modern mammals. However, the amnion remains key to amniote development.

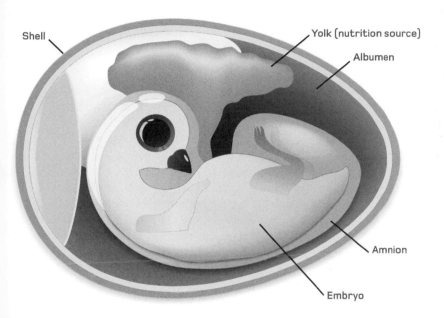

Shell

Yolk (nutrition source)

Albumen

Amnion

Embryo

Amniotes – the group of tetrapods that include reptiles and synapsids – share an amniotic egg. A special membrane – the amnion – surrounds the embryo and keeps it contained within amniotic fluid.

Synapsids rule

Permian life on land was dominated by the Synapsida, a huge amniote group that included the ancestors of mammals. Synapsids were a significant ecological force before the rise of the archosaurs (see page 290). The synapsid fossil record is good enough to give us a useful view of evolutionary transformations relevant to mammal anatomy and biology. A key feature differentiating synapsids from other amniotes is the presence of a single skull opening behind the eye socket.

Synapsids originated during the Carboniferous, from an ancestor that also gave rise to reptiles. Early synapsids – examples include the sail-backed *Dimetrodon* – were reptile-like overall. These early synapsids are informally termed 'pelycosaurs'. Around 300 mya, during the Late Carboniferous, *Dimetrodon*-like synapsids gave rise to a more mammal-like group: the therapsids. Increasingly mammal-like proportions, tooth anatomy and gait evolved within this group; it is likely that the members of one therapsid group – the cynodonts – were furry and broadly mammal-like overall.

Sail-backed *Dimetrodon* is the most famous of early synapsids. The long spines on its back were partially connected by webbing, the resulting sail perhaps having a role in display.

The end-Permian event

One of the biggest extinction events in the history of life – possibly the biggest of all – occurred 250 mya, at the end of the Permian. It seems that several separate extinction pulses occurred, though whether all were connected remains unknown. Some studies indicate that as many as 70 per cent of land-living organisms and over 95 per cent of marine ones died at this time.

The cause of the event remains debated. During Late Permian times, the continents were united, forming the supercontinent Pangaea. The relatively small coastline meant that much of the continental interior was arid and incredibly hot, and it may be that environmental stress ultimately caused the extinction of many things. It has also been suggested that major volcanic eruptions or a release of methane from the sea floor resulted in, or contributed to, climatic warming that caused the extinction. The event saw a substantial reduction in synapsid diversity, the loss of some marine invertebrate groups and a reduction in insects.

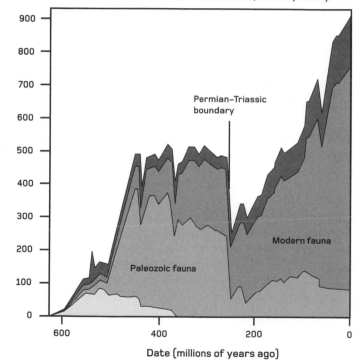

Graphs that chart the diversity of species over geological time depict an obvious drop coinciding with the massive end-Permian extinction event. Life did recover, but only slowly.

The origin of mammals

After evolving during the Late Permian, non-mammalian cynodonts underwent many changes that included becoming smaller and shorter-tailed and evolving a more erect gait. By the Late Triassic, around 210 mya, mammals had evolved from these cynodonts and shared a set of anatomical, behavioural and physiological features with them.

There is some disagreement over the terminology used for these animals. Experts generally term them 'early mammals'. However, some specialists argue that the term 'mammal' should be restricted to the group that includes the living groups alone (monotremes, marsupials and placentals), in which case the mammal-like cynodonts outside this group are termed 'mammaliaforms'. Many Mesozoic mammals were small, furtive insectivores that must have resembled shrews or mice. Recent discoveries have, however, shown that early mammals encompassed more diversity than this. Some were gliders, swimmers, mole-like burrowers or badger-sized predators.

Several early mammals are known from good remains that reveal much about their diets, lifestyles and sensory abilities. *Morganucodon* lived in Wales during the Triassic, and was a small, shrew-like predator.

The ruling reptiles

Archosaurs – the so-called 'ruling reptiles' – include dinosaurs, pterosaurs and crocodylians. Crocodylians survive to the present, but are merely one lineage of what was once a substantially more important, more diverse archosaur group known as the crocodile-line archosaurs. After the disappearance of large synapsids in the end-Permian event, these animals dominated life on land.

Some extinct members of this group may either have been amphibious or terrestrial and predatory – something like modern monitor lizards. But others were markedly different. Aetosaurs were armoured and either omnivorous or herbivorous. Several groups termed rauisuchians were typically predatory, with serrated, blade-like teeth. Some were bipedal, others had dorsal sails; some were toothless and might have been herbivorous. Some rauisuchians were enormous, in cases exceeding 8 m (26 ft) in length. Several body shapes and lifestyles occupied later by dinosaurs initially evolved among these animals.

Saurosuchus is one of the rauisuchians, a group of crocodile-line archosaurs that were the top predators of their day.

The rise of dinosaurs

Archosaurs split early in their history into one lineage that led to crocodiles and their many relatives, and another that led to dinosaurs and their relatives. All early members of this second lineage – termed 'Ornithodira' – were small, long-legged, slender animals, typically less than 1 m (40 in) long. While some were predators, leaf-shaped tooth crowns in others suggest that they were omnivores or herbivores. Pterosaurs emerged from among these animals, as did the Dinosauromorpha group that ultimately gave rise to the dinosaurs.

It seems that all early ornithodirans were secretive animals, lower in the food chain than the larger, more formidable crocodile-line archosaurs of the time. Proper dinosaurs originated around 245 mya, close to the boundary between the Early and Middle Triassic. The oldest members of the group include *Nyasasaurus* from Tanzania and a number from Brazil and Argentina, including *Buriolestes*, *Panphagia* and *Herrerasaurus*. This pattern suggests that dinosaurs originated in the southern continents.

Herrerasaurus is a South American dinosaur with long, recurved teeth and large curved hand claws. It was clearly a predator of other reptiles, large and small.

Turtle origins

Turtles are unmistakeable, their key feature being the shell. Remarkably, this is a modified ribcage covered in horny plates. The fact that the limb girdles are *inside* the ribcage is one of the most incredible evolutionary transformations to have occurred in vertebrate history.

Fossils show that turtles with a modern-style shell were present by the Late Triassic, about 220 mya. They are so unusual that experts have long argued over where they fit within the reptile family tree. Unlike the lizards, archosaurs and kin of the Diapsida group, they lack openings in the back of the skull. However, genetics and fossils show that turtles are modified diapsids in which those openings are closed. *Eunotosaurus*, from the Middle Permian of South Africa, was a small desert-dweller. It didn't have a shell but its massive flattened ribs give it the appearance of an animal close to the ancestry of shelled forms. It seems to have been a desert-dwelling burrower, which suggests that the modified turtle ribcage initially evolved as a burrowing device.

Giant sea-going turtles, equipped with paddles and a reduced shell with weight-saving gaps, evolved during the Cretaceous. They include *Archelon*, shown here.

Frogs, salamanders and caecilians

Frogs, salamanders and worm-like caecilians form Amphibia, a group known for permeable skins and life histories that involve metamorphosis from aquatic larvae. There are more than 7,000 amphibian species today, the majority of which are tropical frogs.

Many fossil tetrapods are more similar to living amphibians than they are to amniotes, and there is some debate as to how amphibians arose. It seems most likely that living amphibians are miniaturized temnospondyls (see page 280). It is also possibile that caecilians do not share an ancestry with frogs and salamanders. All three living amphibian groups had appeared by the start of the Jurassic. *Triadobatrachus* from the Early Triassic of Madagascar is the oldest-known member of the frog lineage. *Eocaecilia* from the Early Jurassic resembles living caecilians overall, but has four short limbs. Early Jurassic salamanders have yet to be discovered, but we predict that they must have existed. The oldest known so far are from the Middle Jurassic.

Frogs similar overall to living species have been in existence for about 200 million years. Most frogs from the Mesozoic were small amphibious animals less than 10 cm (4 in) long.

Marine reptiles

The seas of the Mesozoic were inhabited by numerous aquatic reptiles, equipped with paddles, tail fins and streamlined bodies. Some were giants of 10–20 m (33–66 ft). Numerous details prove that these animals were part of the same group as lizards and crocodylians, termed Diapsida. Several groups took to marine life during the Triassic, thriving in the warm, shallow seas that covered much of Europe and southern Asia at this time. An ability to give birth to live babies evolved, as did especially big eyes and other features relevant to catching prey in the water.

Sauropterygians included mollusc-eating, sometimes turtle-like placodonts, long-jawed nothosaurs and paddle-limbed plesiosaurs. Ichthyosaurs, the best known of which were shark-shaped reptiles of the Jurassic and Cretaceous, were also important. During the Cretaceous, a group of seagoing lizards – the mosasaurs – evolved, and a new turtle group took to marine life as well. Excepting the turtles, all of these groups perished at the end of the Cretaceous.

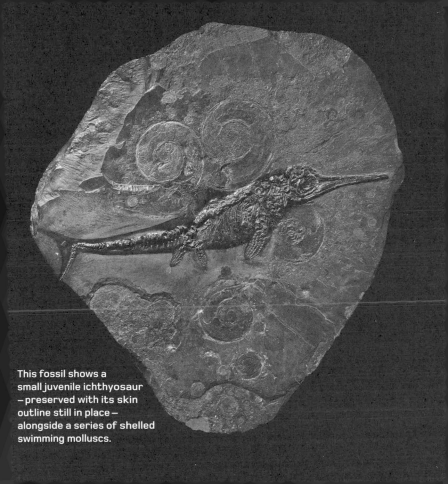

This fossil shows a small juvenile ichthyosaur — preserved with its skin outline still in place — alongside a series of shelled swimming molluscs.

Pterosaurs take wing

The skies of the Mesozoic were inhabited by a group of reptiles with furry bodies that flew with membranous wings. These were the pterosaurs, the oldest fossils of which date to the Late Triassic, about 210 mya. Early pterosaurs were small, with wingspans of less than 1 m (40 in). They were true pterosaurs, however, with all the characteristic features for this group. These include a substantially enlarged fourth finger and a long, slender bone that projected from the wrist towards the shoulder. Both structures served to support the skin membranes that formed the wings.

We have yet to discover a fossil that looks like a 'proto-pterosaur', so the earliest stages of pterosaur history remain poorly known. Anatomical details indicate that pterosaurs are archosaurs – that is, part of the same reptile group as crocodylians and dinosaurs. Within this group, pterosaurs share an elongated neck and a set of other features with dinosaurs and their relatives, so it seems that pterosaurs and dinosaurs are close cousins.

Pterosaurs of many shapes and sizes inhabited the skies of the Mesozoic. Many species had bony head crests, some of which had soft extensions that made them even larger.

The age of dinosaurs

During the Late Triassic (240–200 mya), a wave of extinction events led to the disappearance of most of the previously dominant crocodile-line archosaurs. A group of rare, small, lightly built archosaurs that had already been in existence for about 30 million years finally had the opportunity to take advantage. These were the dinosaurs. Before the Triassic was over they had come to dominate life on land, attaining body sizes well exceeding those of any land-living animals that had evolved beforehand.

Early in their history, dinosaurs split into three major groups, all of which went on to evolve giant size and phenomenal diversity in anatomy and behaviour. All started as small, bipedal predators or omnivores with grabbing hands. Ornithischians evolved a toothless beak in the lower jaw and gave rise to several dynasties of specialized herbivores. Theropods evolved large, curved hand claws and teeth, and jaws suited for predation on other animals. Sauropodomorphs evolved long necks and ever-larger size, eventually becoming quadrupedal and herbivorous.

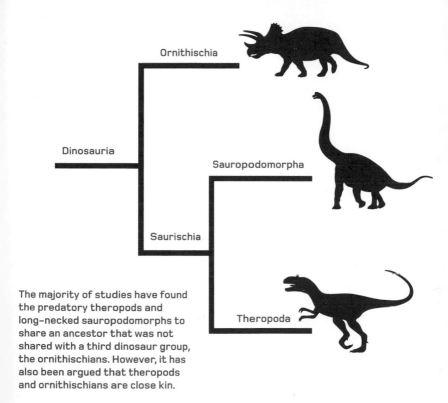

Ornithischia

Dinosauria

Sauropodomorpha

Saurischia

Theropoda

The majority of studies have found the predatory theropods and long-necked sauropodomorphs to share an ancestor that was not shared with a third dinosaur group, the ornithischians. However, it has also been argued that theropods and ornithischians are close kin.

The largest dinosaurs

Of the three dinosaur groups that originated in the Triassic, the sauropodomorphs are associated with giant size. Early sauropodomorphs were small bipeds that weighed less than 10 kg (22 lbs). During the later part of the Triassic, and throughout the Jurassic, they increased in size, soon exceeding 10 tonnes (11 tons). Species weighing 30 tonnes (33 tons) and possibly even 100 tonnes (110 tons) had evolved by the end of the Jurassic. Such giants existed throughout the Cretaceous, right up to its close 66 mya. A long neck, an amount of pneumaticity in the skeleton and huge size were taken to an extreme in sauropods, the most advanced of the sauropodomorphs. Limbs became slender columns and hands converted from clawed, grabbing organs to compact, hoof-like structures. Such features show that sauropods were fully terrestrial, and able to use their flexible necks to reach high up into treetops but feed from ground level too. The size and strength of these animals means that they must have exerted a considerable influence on the landscapes of the time.

Giant sauropods were the largest animals ever to walk on land. Some rivalled large whales in length and mass.

Archaeopteryx

The fully feathered skeletons of *Archaeopteryx* – preserved in the fine-grained lithographic limestone of the Solnhofen region in Germany – are regarded as the most valuable fossils ever discovered (1861). Its modern-looking feathers immediately led to it being identified as an ancient bird. But its toothed jaws, three hand claws and long, bony tail also made it appear transitional between archosaurian reptiles and birds. Here was powerful support for evolution, and the first evidence that early birds lived along their dinosaurian cousins during the Late Jurassic, 150 mya.

Highly similar to theropod dinosaurs like *Velociraptor* and *Troodon*, *Archaeopteryx* was clearly a close relative. Fossils discovered since the 1990s also confirm that dinosaurs of this sort possessed *Archaeopteryx*-like feathering. Although *Archaeopteryx* is no longer the sole key fossil in understanding bird origins, it remains important in terms of the history of palaeontology and in the amount of anatomical detail it preserves.

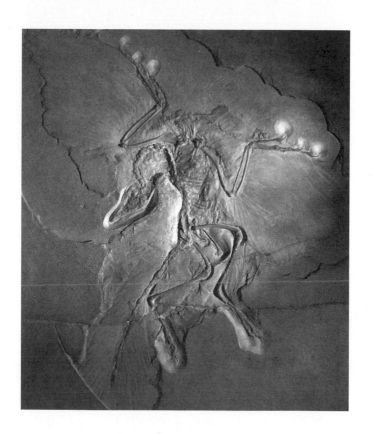

Flowering plants

The most familiar and important modern plant group are the flowering plants – Angiospermae. Water lilies, magnolias, palms, orchids, grasses, oaks, buttercups, sunflowers and hundreds of other groups belong here. There are more than 290,000 species alive today, in over 400 families.

Fossils show that angiosperms did not exist, or were rare, prior to the Early Cretaceous, about 130 mya. Only during this time did the group appear – initially as small, rare, herb-like species, perhaps associated with shallow pools. Several extinct groups look like potential angiosperm ancestors, and they probably evolved from one particular group of gymnosperms – the group that includes cycads and conifers. A number of features make angiosperms unusual: the carpel, a flask-shaped structure that contains the ovary at its base and the stigma at its tip – the site where pollen is received; and a stamen – a long filament that produces pollen at its tip. These structures, surrounded by the modified, often brightly coloured leaves we call petals, form the flower.

Snakes

Snakes are an unusual group of lizards that evolved during the Middle Jurassic, about 170 mya. While fossil snakes share many features with modern snakes – limblessness, a lack of mobile eyelids and a long body formed of more than 200 vertebrae – Jurassic and Cretaceous species had limbs. Fossils show that the forelimbs were lost first, several Cretaceous snakes possessing small hindlimbs where knees, ankles and toes were still present.

Some limbed snakes were terrestrial burrowers; others were marine swimmers. The exact context in which snakes first evolved – as burrowers or swimmers – remains a subject of discussion among experts, as does exactly which group of lizards gave rise to snakes. They have most frequently been considered close to the group that includes monitor lizards and kin. Fully modern, limbless snakes had evolved by the end of the Cretaceous, about 70 mya. Three main lineages diverged. One led to the poorly known, burrowing blindsnakes, one to the constricting boas and pythons, and the third to the colubroids (see page 350).

Armoured dinosaurs

During the Early Jurassic, ornithischian dinosaurs underwent profound change. One group – the thyreophorans ('shield bearers') – evolved rows of bony nodules along their necks, backs and tails. Early examples looked much like other small, bipedal ornithischians, but with armour – *Scutellosaurus*, for example.

Later thyreophorans, such as *Scelidosaurus*, increased in size, evolved more extensive, elaborate armour and made a transition to walking on all fours. It seems that animals like this gave rise to the two main thyreophoran groups: stegosaurs and ankylosaurs. Stegosaurs were a Jurassic group, famous for the giant, diamond-shaped plates of the Late Jurassic *Stegosaurus*. They survived into the Early Cretaceous before disappearing. Broad-bodied and heavily built ankylosaurs developed horns, sideways-projecting spines and spikes, tail clubs and hip shields, as well as complex chewing motions and peculiar twisting tubes inside the snout. Several ankylosaur groups persisted to the very end of the Cretaceous.

Ankylosaurs were large (up to 7 m/23 ft long), covered in armour plates, and with bony tail clubs and prominent facial horns.

Modern sharks, rays and chimaeras

Chondrichthyans, or cartilaginous fishes, probably originated in the Silurian approximately 440 mya. By late Mesozoic times, archaic chondrichthyans had disappeared and modern lineages had emerged from among the groups of the Permian and Triassic. Species close to living bramble sharks, cookiecutters, sleeper sharks, wobbegongs, nurse sharks, ragged-toothed sharks and goblin sharks were all present by the Late Cretaceous. Other modern groups have fossil records with older appearance dates: horn sharks were present by the Middle Jurassic, for example.

Modern chimaeras were present from the Jurassic onwards. Chimaeras possess large, paired plates in the upper and lower jaws, the tips of which protrude from the mouth to create the appearance of rat- or rabbit-like incisor teeth. Another group of chondrichthyans – the batoids or rays – also proliferated during Jurassic and Cretaceous times after diverging from an ancestor that also gave rise to sharks. Their key hallmarks include winglike pectoral fins and pavements of interlocked, flattened teeth.

Sharks like this great white belong to a group that had origins in the Jurassic or Triassic. Sharks as a whole are an ancient group, although modern species like the great white are not especially old.

The Mesozoic Marine Revolution

Significant changes affected the marine environment throughout the Mesozoic (220–66 mya). New groups of organisms evolved, older ones disappeared and ecosystems changed fundamentally. Known as the Mesozoic Marine Revolution (or MMR), this overturn resulted in the appearance of animal and plant communities that were essentially like modern ones.

The main driving force of this change appears to have been the evolution of new invertebrate groups – gastropods, lobsters and crabs – that were good at breaking into the shells of other organisms. Shell-crushing chondrichthyans and marine reptiles may also have contributed to this phase of predation. Invertebrate groups specialized for grazing on algae also became more important, including gastropods and sea urchins. As a consequence of these changes, burrowing became more common, shells became thicker and more difficult to break, immobile groups disappeared or became rare, and groups able to escape predation by moving increased in number.

Modern seafloor habitats are still dominated by many species that evolved or rose to prominence in the Mesozoic, such as crustaceans.

Flowers and insects

Flowers, such an obvious component of angiosperm plant anatomy and biology (see page 308), clearly co-evolved with insects. They use tricks to attract insects and get them to collect and transfer pollen; they frequently possess colours and patterns specifically designed for the UV-sensitive insect eye.

Flowering angiosperms appeared during the Early Cretaceous (about 130 mya) and it has long been predicted that their appearance coincided with new evolutionary developments among insects. Some evidence supports this view: both fossil and molecular data show that plant-eating beetles diversified in step with angiosperms, and that bees, pollen-eating wasps, butterflies and moths all probably evolved at the same time. Other insect groups that use flowers – sawflies, craneflies and fungus gnats – were present before angiosperms appeared, however, and do not seem to have undergone any new burst of evolution once flowers were on the scene. By the Late Cretaceous, some insect groups, such as bees and butterflies, were highly specialized flower-users.

Spiny-finned fishes

Acanthomorphs, or spiny-finned fishes, belong to a group called teleosts, which in turn belong to actinopterygians, the ray-finned bony fishes. With more than 16,000 living fish species, the acanthomorphs have been the dominant fish group since the early Cenozoic, and are the most abundant fishes today.

Oarfishes and kin, cod and kin, gobies and kin, pipefishes, and a group that includes butterflyfishes, perches, sticklebacks, flatfishes, anglerfishes, wrasses and cichlids form the major lineages within this radiation. Fossils show that acanthomorphs of most lineages had appeared by the Late Cretaceous, and the oldest members of the group date to the middle part of the Cretaceous, about 100 mya. Those earliest kinds – they include *Muhichthys* from Mexico and *Xenyllion* from the United States – were deep-bodied, short-headed fishes of shallow, tropical marine environments. Many later acanthomorphs were broadly similar to this, but some took to living in deepwater marine environments, while others came to dominate estuaries, lakes and rivers.

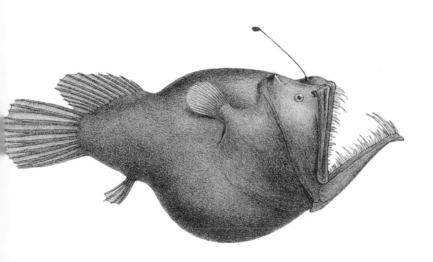

Acanthomorph fishes of several groups colonized the
deep ocean during the past 50 million years. Among
these are the remarkable deep-sea anglerfishes.

Feathers and early birds

Thanks to fossils like *Archaeopteryx* (see page 306), we know that feathers have existed for 150 million years. Scientists long linked their evolution to flight, but we now know that complex feathers were present in dinosaurs before flight evolved, so feathers originated in some other context: they provide insulation, waterproofing, camouflage and protection, are used in display, steering and provide balance and thrust when running or leaping.

Several new bird groups evolved from *Archaeopteryx*-like ancestors during the Late Jurassic. Among the most important were the enantiornithines of the Cretaceous, sometimes called 'opposite birds', since parts of their skeletons had a configuration very different from those of modern birds. Unlike *Archaeopteryx*, enantiornithines had a short tail skeleton. Most had teeth, but elongated jaws and toothlessness evolved. Wire-like or streamer-like feathers were present on the tails and hindlimbs of some species. Enantiornithines were not ancestral to living birds but many of their features presaged those present in birds of today.

A huge number of bird species evolved during the Cretaceous. Like the one shown here, many belong to the group known as the 'opposite birds'.

Giant pterosaurs

During the Cretaceous, groups of membranous-winged, flying pterosaurs were present worldwide, and a few of them contained giants. The most familiar giant was *Pteranodon*, unique to the shallow sea that covered much of North America during Late Cretaceous times, and with wingspans up to 7 m (23 ft).

The azhdarchids also inhabited the Cretaceous world – long-jawed, long-necked and toothless pterosaurs with wingspans up to 10 m (33 ft). The largest flying animals ever, a giant azhdarchid would have weighed over 220 kg (485 lb) and been similar in standing height to a giraffe. Distributed globally during the Late Cretaceous, their fossils show that they occurred in deserts, wooded regions and coastal plains. Many places were inhabited by several species at the same time, these differing in size and presumably in foraging style and diet. The jaw shapes; slender, erect arms and legs; and small, compact feet show that these pterosaurs most likely patrolled terrestrial environments, reaching down to grab small animals from the ground.

Pteranodon is a large North American pterosaur that lived like a modern albatross, soaring over the waves and alighting on the water to catch marine animals.

Duck-billed dinosaurs

The hadrosaurs, or duck-billed dinosaurs, were one of the most abundant and successful Cretaceous dinosaur groups – a group of ornithischians that originated approximately 130 mya, probably in eastern Asia. From here, they colonized the island archipelagos of Europe and the Americas.

A key innovation was the enlargement and development of tooth batteries. Hundreds of interlocked teeth formed blocks at the back of both the upper and lower jaws. The inner surfaces of these blocks were rasp-like while their cutting edges formed long, slicing blades. The name 'duck-bill' refers to the broad, flaring shape of the bones that form the tip of the hadrosaur upper jaw. Head crests were common in hadrosaurs and included solid, spike-shaped crests, arched nasal structures, and hollow, plate-shaped and tubular crests. Debate continues as to which selective forces drove the evolution of these structures, but it is agreed that they were used in visual display and were perhaps brightly coloured and/or boldly patterned.

Many hadrosaurs evolved prominent bony head crests. *Parasaurolophus*, shown here, possessed a hollow crest that housed a series of internal tubes.

Giant horned dinosaurs

The Late Cretaceous ceratopsians, or horned dinosaurs, descended from small, bipedal ancestors that also gave rise to pachycephalosaurs (dome-headed ornithischians). Ceratopsians of the first half of the Cretaceous were mostly bipedal, beak-jawed herbivores, some 1–2 m (3–6½ ft) long.

Towards the middle of the Cretaceous, members of one ceratopsian lineage took to quadrupedality, increased in size and robustness, and evolved brow and nasal horns and a long bony frill. Individuals are often found preserved together, indicating that these were herd-living, social animals. The diverse, elaborate nature of the frills and horns suggests that the primary role of these structures was display. Indeed, they increased in complexity and diversity over time, whereas some lineages reduced or even lost their horns altogether as other parts of their skulls became more elaborate. The last ceratopsians were the largest. *Triceratops* of western North America exceeded 8 m (26 ft) and 10 tonnes (11 tons).

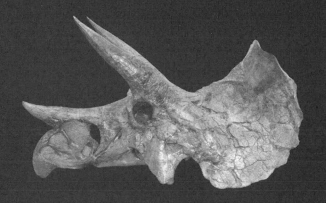

Triceratops (above) and *Torosaurus* (below) were the last and largest of the ceratopsians. They possessed giant hooked beaks and batteries of slicing teeth.

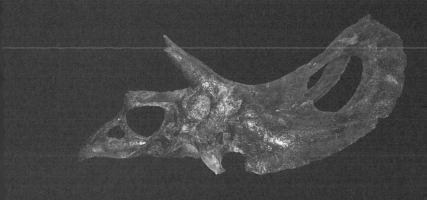

The great tyrant dinosaurs

Theropod dinosaurs were the dominant terrestrial predators of the Jurassic and Cretaceous. Numerous lineages evolved, among the last of which were the giant tyrannosaurs, or tyrant dinosaurs, large species of which exceeded 6 m (20 ft). Fossils show that tyrannosaurs originated during the Middle Jurassic (approximately 170 mya) at small body size, the species of this time being human-sized.

Tyrannosaurs had become giants by the Late Cretaceous, but only after other giant theropod groups, including megalosaurs and allosaurs, had disappeared following a series of extinction events. Proportionally small forelimbs, an especially muscular neck and a very heavily built skull allowing a deadly powerful bite were all typical tyrannosaur features. Species belonging to several lineages became the top predators of North American and Asian dinosaur communities. The best known giant tyrannosaur is, of course, *Tyrannosaurus* of western North America, a super-predator of 12 m (39 ft) and 10 tonnes (11 tons).

Giants like *Tyrannosaurus* possessed thickened skull bones and spike-like teeth. These features allowed them to break bones and kill the largest of prey animals.

The KPg event

The end of the Cretaceous, 66 mya, saw the most famous extinction event of all time. Non-bird dinosaurs disappeared, as did pterosaurs, the great marine reptiles and numerous other groups. This is the KPg event, 'K' being the geological symbol for Cretaceous and 'Pg' for Paleogene. The event marked the end of the Mesozoic and the beginning of the Cenozoic. Many experts argue that the impact of an asteroid or comet caused this event. Traces of the metal iridium, blobs of molten rock and fractured fragments of quartz all provide evidence. The 'smoking gun' is the Chicxulub crater, more than 180 km (112 miles) wide, discovered deep in the rocks of the Yucatán Peninsula, Mexico. It is of exactly the right age. Some experts think there were additional reasons for the extinction. Many Cretaceous groups were in decline before the event and were, therefore, vulnerable. A prolonged period of volcanic activity may have disrupted ecosystems of the time, and a major drop in sea level also seems to have had negative ecological impacts. Had conditions improved, the affected groups would likely have recovered. But things did not improve.

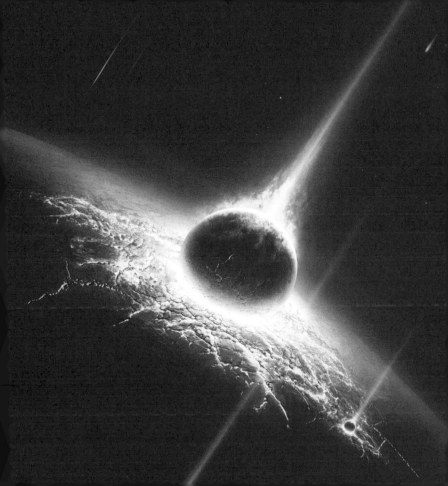

After the dinosaurs

The KPg event disrupted ecosystems worldwide, resulting in impoverished and partially empty habitats. Fossil evidence shows that life responds relatively quickly after extinction events, although unbalanced ecosystems exist for a while. In the seas, several plankton groups built up large populations within a few hundred thousand years; on land, ferns saw a brief boom.

With non-bird dinosaurs gone, lizards, birds and mammals were no longer under the same predation pressures that they had been. Birds and mammals underwent a burst of diversification, the majority of modern lineages evolving within just 10 million years of the Cretaceous. Some species became giants relative to their ancestors – during the Late Paleocene (60 mya) there were flightless birds, mammals and snakes weighing over 50 kg (110 lb). Lizards, snakes, turtles and crocodylians would all remain important and abundant, at least in tropical and subtropical environments. However, the stage had been set for a world in which life on land was dominated by birds and mammals.

Large birds were important in many environments after the end-Cretaceous extinction. The flightless *Gastornis*, a forest-dwelling plant-eater, stood about 2 m (80 in) tall.

The global greenhouse

The Paleocene – the first 10 million years after the Cretaceous – and subsequent Eocene, lasting 23 million years, saw increasingly hot temperatures. Termed the Paleocene-Eocene Thermal Maximum (PETM), at its peak the average global temperature was approximately 23°C (74°F). Oceans were warm and tropical forests stretched from pole to pole. Giant reptiles were abundant. Big herbivorous lizards and freshwater turtles, large crocodylians and gargantuan snakes inhabited South America, Asia and elsewhere. The most famous – *Titanoboa*, a predatory snake from Colombia – is estimated to have been 12 m (39 ft) and 1,000 kg (2,200 lb). Why did conditions become so warm? Data recorded in sediments and fossils show that carbon concentrations increased markedly, apparently as a consequence of methane release from sea-floor sediment. This presumably happened as a result of a change in ocean circulation or chemistry, the ultimate explanation perhaps being the influx of huge quantities of freshwater from a previously contained giant lake in the Arctic.

Paleocene South America was home to the largest snake of all time, *Titanoboa*. Large crocodylians and turtles shared its hot, humid habitat.

Grasses and grasslands

Grasslands, plains, prairies and steppes are dominated by a single group of flowering plants: grasses. Bamboos and rice, barley and wheat are all grasses. A recent evolutionary event, grasslands appeared on most continents between about 20 and 10 mya, mostly during the Miocene. South America is unusual in that it had grasslands 30 mya, a fact that made a major difference to the history of mammal evolution there.

Grasses do not rely on insects for pollination, instead releasing their pollen into the wind. They are also highly resistant to grazing and fire, and are able to recover quickly from extensive damage. After the decline of the global rainforests that happened during the Oligocene and Miocene, grasses spread rapidly across non-forested areas. They proved hugely important in the evolution of mammals. Horses, rhinos and antelopes evolved most of their key features on plains, as did fast-running, keen-eyed predators, such as dogs and cheetahs. Features important in the evolution of our group – the hominids – also evolved following life on the plains.

Bats take wing

Bats form one of the largest mammal groups, with more than 1,200 species occurring worldwide today. They are the only mammals to have achieved powered flight. Fossil remains – mostly jaw fragments – are abundant in cave deposits and show that bats of some living groups have been present since the Oligocene. Older fossils include *Icaronycteris* from the Eocene of the United States; it does not belong to bat groups that survive today, but appears to have used echolocation to hunt moths.

One of the most remarkable fossils is the so-called 20-claw bat – also from the Eocene of the United States – officially named *Onychonycteris*. In contrast to all other bats, *Onychonycteris* had five claws on each wing, a feature that makes it the most structurally 'primitive' bat we know. Proportionally, its hindlimbs were longer and its forelimbs shorter than those of modern bats, while the anatomy of structures in its ears and throat make it difficult to determine whether it was capable of echolocation or not. Its well developed wings certainly show that it was a true flier.

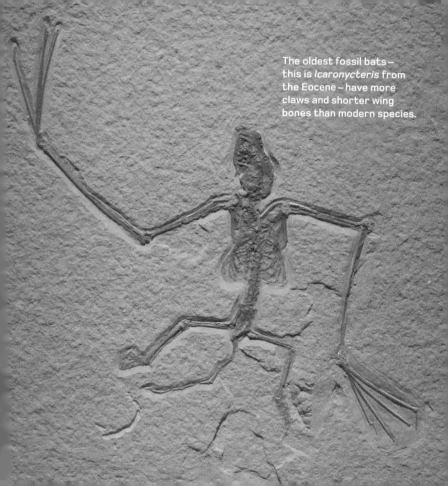

The oldest fossil bats — this is *Icaronycteris* from the Eocene — have more claws and shorter wing bones than modern species.

Modern birds arise

Feathered dinosaurs that we term 'birds' originated during the Jurassic and several groups had evolved by the Cretaceous. Only one of these survived beyond it – the small, toothless neornithines. With enlarged brains and expanded bills, these are 'less dinosaur-like' than other bird groups.

Early in their history, neornithines diverged into palaeognaths and neognaths, reflecting differences in palatal structure. Palaeognaths include the flightless ostriches and kin; neognaths include over 10,000 living species. Wildfowl, gamebirds, cuckoos, seabirds, waders, hawks, owls, parrots, pigeons and perching birds (the huge group that includes crows, thrushes, warblers, sparrows, finches and so on) are all neognaths. Excellent fossils from Messel in Germany and Wyoming in the United States show that early ducks, waders, swifts, hawks and others existed by the Eocene. This suggests that a rapid diversification of tens of modern lineages must have occurred both during the Late Cretaceous and just after it.

The feathery outlines preserved on some fossil birds, such as this *Scaniacypselus*, show that body and wing shapes similar to those present today evolved during the Eocene.

Fishes and the deep-sea realm

Popular accounts of evolutionary history typically discuss fish as a prelude to the evolution of tetrapods in the Devonian, but then fail to revisit them. In reality, fish of many groups underwent extraordinary change after that time, and some of the most impactful events in fish history occurred within the last 70 million years. This is especially true for actinopterygians or ray-finned fishes, and in particular for the subgroup termed the 'teleosts'.

Teleost genetics show that the group exploded in diversity close to the end of the Cretaceous, the vast majority of modern lineages originating at this time. Fossils support this view. Complete fish fossils from Bolca, in Italy, show that modern reef-dwelling teleosts were present by about 35 mya, as were open-ocean and freshwater groups. A few had also adapted to life in an environment that actinopterygians had not visited before – namely, the deep sea. Viperfishes, bristlemouths, hatchetfishes and other weird deep-sea teleosts had evolved by 25 mya, all represented by a reasonably good fossil record.

Modern-looking fish, such as this moonfish, had evolved by about 40 mya. Moonfish are unique to the Indopacific today but once occurred in shallow seas that covered Europe.

Mammals of the Miocene

Cenozoic vertebrate life on land was dominated by mammals. The hot, forested world of the Paleocene and Eocene saw mammals diversify into tens of lineages, many of which survive today. But the cooling and drying that occurred during the Oligocene and Miocene (25–5 mya) led to the disappearance of many groups, and to evolutionary change in those that remained.

The cooling and drying led to the appearance of great grasslands. Numerous herbivorous groups took to life on the plains, each evolving longer legs, larger size and an improved ability to feed on the tough leaves of the grasses. By about 18 mya, during the Miocene, habitats worldwide were occupied by pigs, peccaries, camels, antelopes, cattle, horses and rhinos. It was the golden age of mammals. Rodents – rats, squirrels, beavers and kin – also moved into these new environments. Many became specialized consumers of seeds, and good burrowers and tunnel builders. This proliferation of new animals led to a burst of evolution among predatory mammals too.

During the Miocene, North America was home to numerous horses and camels, as well as to less familiar animals like the clawed chalicotheres.

The rise of cats

One of the most important groups of placental mammals – 'important' in the sense that they have had a major influence on the evolution of many other groups – are the carnivores, or carnivorans. Carnivorans appeared about 55 mya, as small, tree-climbing predators of tropical forests; by about 40 mya, they had given rise to dogs, bears, seals, weasels, civets, hyaenas and cats. Carnivorans are mostly predators of other vertebrates, although insect-eating, fruit-eating and leaf-eating types evolved within the group as well. Cats are the most specialized terrestrial carnivorans and are more committed to a carnivorous diet than any other group. They possess a reduced number of teeth and have a body shape suited for the stealthy ambush of prey. Cats began their history with conical canine teeth, but enlarged, flattened upper canines, sometimes with serrated edges, evolved in the sabretooths. Modern cats arose in the Late Miocene – about 11 mya – and split into the lineages leading to big cats, pumas and cheetahs, and the small cats of the Americas, Africa, Europe and Asia.

The so-called 'sabretoothed' design evolved several times in the history of cats. The most famous sabretooth is the giant *Smilodon*, shown here.

Venomous snakes

The Cenozoic is so often termed the 'Age of Mammals' that it is easy to forget that other groups diversified massively during this time as well, among them birds, bony fishes, frogs and snakes. The majority of snakes belong to the group Colubroidea, more than 2,500 species of which exist today. While many are non- or mildly venomous, some are highly venomous, using hollow, erectile fangs at the front of the mouth to inject venom into other animals, or to squirt venom into the eyes of attackers.

Fossils show that highly venomous colubroids are newcomers on the scene, probably originating about 23 mya. In fact, molecular data shows that some events within venomous snake evolution – the evolution of sea snakes, for example – mostly occurred within the last five million years. Some aspects of snake evolution were driven by co-evolution with mammals, either because the snakes became specialized predators of mammals (mostly rodents), or because they now had to defend themselves from groups like carnivorans and primates.

Colubroids, such as this mangrove snake from tropical Asia, belong to an enormous, globally distributed, highly successful snake group.

The spread of perching birds

Of the 10,000 species of birds that exist today, some 6,000 belong to a single group: the perching birds, properly called 'passeriforms' or 'passerines'. They are, of course, not the only birds capable of perching, but their generally small size, adaptable diets, and feet with an enlarged, opposable first toe are among the keys to their success. Crows, warblers, thrushes, swallows, larks, flycatchers, finches and sparrows are all passerines.

Genetics and anatomy show that birds unique to New Zealand – New Zealand wrens – are one of the oldest passerine groups. Several others, mostly associated with South America, Africa, tropical Asia and Australasia, evolved next. Finally, the globally distributed oscines appeared – the group that includes crows and sparrows. The pattern shows that passerines were originally birds of the southern continents, which later staged several northward invasions. The oldest oscine fossils are from Australia, showing that even they likely originated there before spreading north to Africa, Madagascar and Asia, and then Europe and the Americas.

Sparrows possess strong bills that can break open seeds as well as grab insects. They are part of a huge group of birds termed the 'passerines'.

Elephants, mastodons and mammoths

Elephants are familiar modern mammals, but are merely the only living members of a once more diverse group, termed 'proboscideans'. The oldest fossil proboscideans were similar in size and proportion to pigs or tapirs and lacked the unusual nasal regions, jaws and faces of later kinds.

Fossils show how animals like this gave rise to a lineage that evolved larger size, column-like limbs and a widened, deepened nasal region linked with the most remarkable proboscidean innovation: the trunk, a structure that allowed these short-necked animals to reach the ground and even up into trees. Long, protruding incisor teeth – the tusks – served as tools in fighting, foraging and digging. All of these early proboscideans were restricted to the isolated continent of Afro-Arabia until it collided with Eurasia about 22 mya. They then spread across Eurasia, this collision being termed the Proboscidean Datum Event as a consequence. Several proboscidean groups – namely mastodons and mammoths – also entered the Americas.

After originating in Africa, proboscideans spread to Asia, Europe and the Americas. The American mastodon — shown here — is one of many species that once inhabited North America.

The primates

The primates are of special relevance to the story of human evolution. In addition to humans, primates include apes, monkeys, tarsiers, the Madagascan lemurs and the African and Asian bushbabies and lorises. Several features make primates unusual: the thumb and the hallux (or big toe) are both enlarged and oppose the other digits; nails are present instead of claws; and the eyes are large and face forwards.

Virtually all of these features evolved in step with a tree-dwelling lifestyle where flexible fingers and toes grasped at branches and twigs. A diet involving insects, fruits and flowers probably explains why primates evolved excellent colour vision, large, forward-pointing eyes and an accurate ability to judge distance. Being small, they were in danger from large snakes and predatory birds and mammals. Some experts suggest that this predation may have driven some aspects of early primate behaviour and biology. The long hindlimbs and good leaping abilities of many primates, for example, plausibly originated as anti-predator features.

Among the most famous of primate groups are the lemurs. Around 100 lemur species inhabit Madagascar. The most familiar of them is the highly social, ring-tailed lemur.

Monkeys and other anthropoids

During the Oligocene (40 mya), a new group of primates evolved from lemur-like ancestors. Termed 'anthropoids', they had larger brains and flatter faces than earlier primates. Many early anthropoids were small, with robust jaws and teeth suited for eating nuts and seeds; they inhabited both southern Asia and northern Africa. Several lineages evolved from these. Virtually all of these primates are considered monkeys, although the term is most frequently applied to the members of two surviving lineages: platyrrhines and cercopithecoids.

Originating in Africa, platyrrhines successfully crossed the Atlantic at some point, presumably by floating on giant masses of vegetation. Spider monkeys, capuchins and marmosets are all members of this group. Cercopithecoids diversified in Africa, Europe and Asia to give rise to colobus monkeys, macaques and baboons. One monkey-like lineage, closely related to cercopithecoids, evolved larger size, reduced and lost their tails, and gave rise to the anthropoids we call apes.

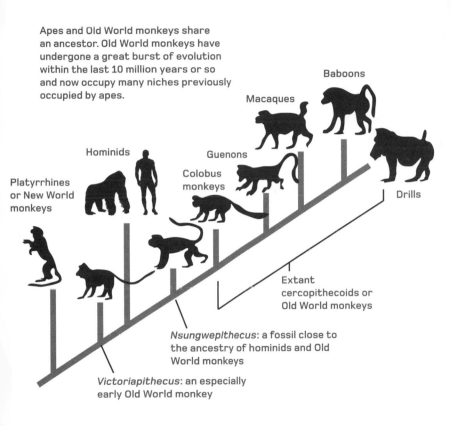

Apes and Old World monkeys share an ancestor. Old World monkeys have undergone a great burst of evolution within the last 10 million years or so and now occupy many niches previously occupied by apes.

Baboons

Macaques

Hominids

Guenons

Colobus monkeys

Platyrrhines or New World monkeys

Drills

Extant cercopithecoids or Old World monkeys

Nsungwepithecus: a fossil close to the ancestry of hominids and Old World monkeys

Victoriapithecus: an especially early Old World monkey

Planet of the apes

Today – with the exception of humans – apes are restricted in range. Gibbons are long-armed canopy-dwellers of southeastern Asia, orangutans are restricted to Borneo and Sumatra, and chimps, bonobos and gorillas inhabit tropical and subtropical Africa. Fossils show that apes – collectively termed 'hominoids' – were once far more diverse in anatomy, biology and distribution, their range encompassing Europe, the Middle East, southern Asia and Africa. More than 50 species existed during the Miocene (23–5 mya) and some experts refer to Earth of the time as the 'planet of the apes'. Many Miocene apes were small compared to modern ones, with different proportions. Some were monkey-like and occupied lifestyles associated with monkeys today. Large apes similar to modern species evolved about 10 mya when skull and body shapes more like those of orangutans, gorillas and chimps evolved. Some of these animals became much larger than earlier species, the most famous example of this being the Asian *Gigantopithecus*. Its teeth and jaws suggest that it was 3 m (10 ft) tall and weighed over 500 kg (1,100 lb).

Many extinct **apes** looked different from the modern species. *Gigantopithecus*, shown here, appears to have been unusually large and strongly built.

The Hominins

Traditionally, the primate lineage that includes humans was regarded as a distinct family (Hominidae) that split from apes (Pongidae) more than 20 mya. But this view does not reflect the evolutionary closeness between humans and some nonhuman apes, specifically chimpanzees. Fossil and DNA evidence shows that the human and chimp lineages separated very recently, perhaps just five or six mya. Clearly, the human lineage is a recently evolved subset of the ape family, and specifically of the African ape group.

Today, experts therefore regard apes and humans as part of the same group: the name 'Hominidae' is preferred since it is older than Pongidae. Within Hominidae, African apes (gorillas, chimps and humans) are the Homininae or hominines. Within Homininae, the human lineage is Hominini, or the hominins. The vast majority of hominins are extinct. The group includes species similar to *Homo sapiens* – like Neanderthals and *Homo erectus* – as well as the somewhat more chimp-like australopithecines.

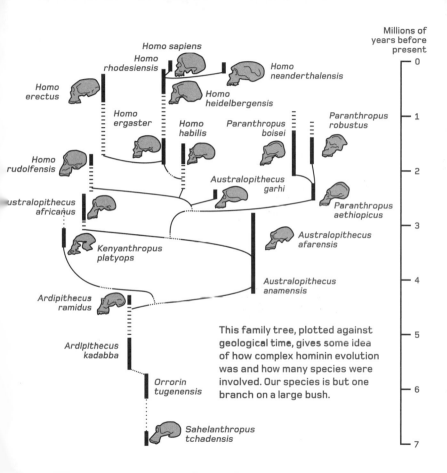

Millions of years before present

Homo sapiens

Homo rhodesiensis

Homo neanderthalensis

Homo erectus

Homo heidelbergensis

Homo ergaster

Homo habilis

Paranthropus boisei

Paranthropus robustus

Homo rudolfensis

Australopithecus garhi

ustralopithecus africanus

Paranthropus aethiopicus

Kenyanthropus platyops

Australopithecus afarensis

Ardipithecus ramidus

Australopithecus anamensis

Ardipithecus kadabba

This family tree, plotted against geological time, gives some idea of how complex hominin evolution was and how many species were involved. Our species is but one branch on a large bush.

Orrorin tugenensis

Sahelanthropus tchadensis

Brains and bipedality

Hominin apes originated as tree dwellers, and their reliance on trees whose locations and fruiting schedules had to be recalled encouraged evolution of relatively large brains from early in their history. Another key human characteristic, bipedal walking, no longer seems quite so closely linked to brain size as was once suspected. It has been suggested that chimpanzee-like hominins became bipedal because two-leggedness gave them advantages in terms of carrying objects, seeing further, speed or endurance.

Brain size has increased three-fold during human evolution (from c. 350 cm³ to over 1300 cm³). The brain surface has also become more intensely folded, allowing for growth of neurons and greater 'storage space'. Why our large, complex brains evolved remains a subject of debate. Perhaps an increase in social interactions drove development, perhaps a need to perform complex tasks was key, or perhaps intelligence was sexually selected (see page 66). It may also be that hominin brain size was driven by constant evolutionary pressure to adapt to new situations and resources.

At some point during ape history, members of one group developed bipedal habits. It remains controversial as to whether knuckle-walking apes evolved from these bipeds, or vice versa.

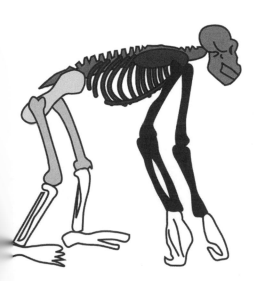

Lucy and other australopithecines

In 1925, Raymond Dart described the skull of a juvenile *Australopithecus* from Taung, South Africa. Since then, at least ten australopithecine species have been named, showing that these animals lived across southern and eastern Africa. One of the most complete specimens was found in 1974 at Hadar, Ethiopia, by Donald Johanson and colleagues. They dubbed it 'Lucy' after 'Lucy in the Sky with Diamonds' by The Beatles. Lucy lived about three mya and would have been 1.1 m (3²/₃ ft) tall. The shape of the animal's pelvis suggests that it was female, but this is disputed.

How australopithecines were proportioned and how they walked and lived are all subject to debate. Some experts point to evidence for human-like walking and a life partly spent on the savannah. Others argue that australopithecines were woodland creatures, good at climbing, and proportioned more like chimps. Some were powerfully built and with thick jaws and massive teeth. Others were lightly built. The human genus – *Homo* – emerged from among *Australopithecus* around three mya.

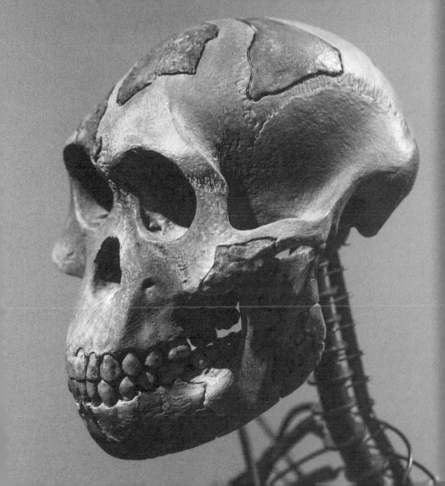

'Handy man' and its relatives

The discovery of australopithecines and of hominins such as *Homo erectus* and the Neanderthals (see pages 370 and 376) did much to improve knowledge of human evolution. But a gap in the fossil record existed: namely that involving species between australopithecines and human-like *Homo*. This gap was filled in 1964, when Louis Leakey and colleagues named *Homo habilis*, discovered in 2.1-million-year-old rocks at Olduvai Gorge, Tanzania. Its face was less protruding than that of an australopithecine, yet less flat than that of hominins like *Homo ergaster*. Its limbs were mid-length between australopithecines and modern humans. And its brain was larger than that of an australopithecine, but smaller than that of *Homo ergaster*. Stone tools indicated that it was a tool-user, hence its name ('handy man'). While the name *Homo habilis* remains widely used today, it has been suggested that the primitive nature of this species compared to other *Homo* species makes this classification questionable. Perhaps, some say, it would be more appropriate to regard *Homo habilis* as a species of australopithecine instead.

Several of the extinct hominins would have looked something like a cross between a chimp and a human. This reconstruction shows *Homo habilis*.

Homo erectus and kin

In 1891, French scientist Eugène Dubois discovered the partial skull and other bones of a fossil hominin on Java. Dubois named it *Pithecanthropus erectus* ('upright ape man'). Here was a hominin more like modern humans than living apes, yet less like modern humans than Neanderthals. Today we term this species *Homo erectus*. It was a large, tool-using hominin with prominent brows.

Similar fossils were later discovered across Asia, Europe and Africa. The best known include 'Peking Man' from Zhoukoudian, China, and 'Turkana Boy' from Kenya, a skeleton representing a young male. More recently, good specimens have been discovered at Dmanisi, Georgia. Hominins of this sort span an enormous range in time: some are more than 1.5 million years older than the youngest (about 200,000 years old). They might represent several species. Either way, *Homo erectus*-like hominins colonized Europe and Asia after migrating out of Africa, and their anatomical features show that they were more like Neanderthals and our own species than were *Homo habilis* and the australopithecines.

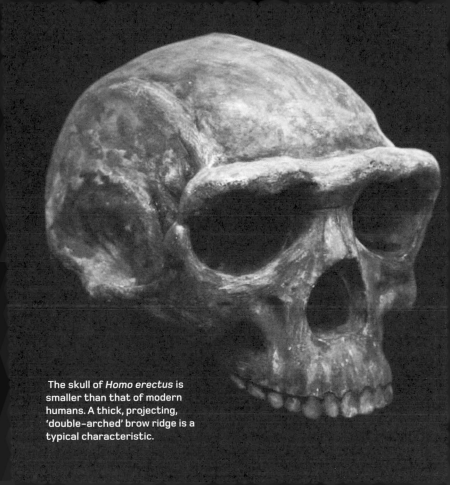

The skull of *Homo erectus* is smaller than that of modern humans. A thick, projecting, 'double-arched' brow ridge is a typical characteristic.

Hobbits and others

The 21st century has seen a burst of discoveries that challenge ideas on hominin diversity and evolution. Among the most famous is the Hobbit – technically, *Homo floresiensis* – a dwarf, island-dwelling hominin discovered at Liang Bua Cave, Flores, Indonesia, in 2003. It was alive until 50,000 years ago.

An idea present throughout Hobbit research is that the fossils are pathological specimens of a known species that suffered from microcephaly (a condition where the head does not develop to normal size). But these arguments are not convincing and do not explain distinctive features seen throughout the Hobbit skeleton. An idea favoured by many hominin fossil experts is that Hobbits were island-dwelling descendants of a larger species, such as *Homo erectus*. A few features of the Hobbit skeleton, however, suggest the possibility that they were not close relatives of *Homo erectus*, but instead belonged to a more archaic hominin lineage. If this is valid, the small size of this species may not have been a special adaptation, but the normal, ancestral condition.

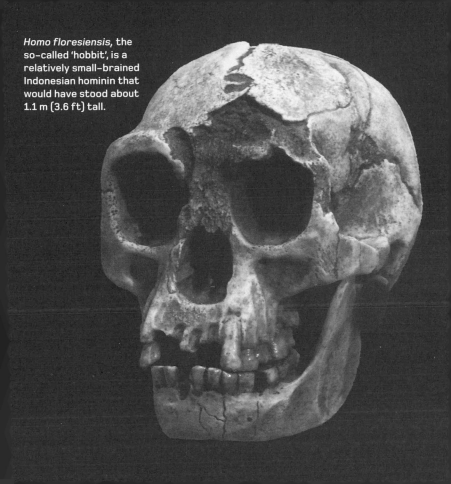

Homo floresiensis, the so-called 'hobbit', is a relatively small-brained Indonesian hominin that would have stood about 1.1 m (3.6 ft) tall.

Language

Primates of many sorts were using vocal signals for millions of years before hominins evolved. But it is only within hominins that the complex, structured method of communication that we term 'language' evolved. Exactly when language originated is difficult to determine, but most experts think that it originated around 300,000 years ago, prior to the time when *Homo sapiens* and Neanderthals diverged from their ancestor.

It is likely that our complex language results from a combination of social intelligence with a memory and understanding of rules and structure and a high degree of control over breathing (and hence of our sound-making apparatus). This combination allows us to generate a complex, variable range of noises, linked with meaning. How and why language originated remains uncertain. Some consider that it may have evolved as hominins developed ways of expressing increasingly complex ideas relevant to hunting, art and social behaviour, and may also have served a role in sexual display.

The Neanderthals

Fossils and genetics indicate that Neanderthals originated about 600,000 years ago from *Homo erectus*-like hominins and were restricted to Ice Age Europe and western Asia. Properly termed *Homo neanderthalensis*, they were heavily built and muscular, but were not short in stature, average heights for males being 1.6 m (5¼ ft). They were probably similar to European populations of *Homo sapiens* in skin colour and hairiness.

Neanderthal culture was complex. They used fire, manufactured tools and clothing and built dwellings from mammoth bones. There is even evidence that they used boats. Neanderthals disappeared about 28,000 years ago when conditions in Europe were especially cold. By this time they had been pushed out of Europe and Asia by the expanding *Homo sapiens* population and were restricted to the Atlantic fringes of the Iberian Peninsula. However, genetic evidence shows that Neanderthals and modern humans interbred, and that Neanderthal genes persist today in many populations of our species.

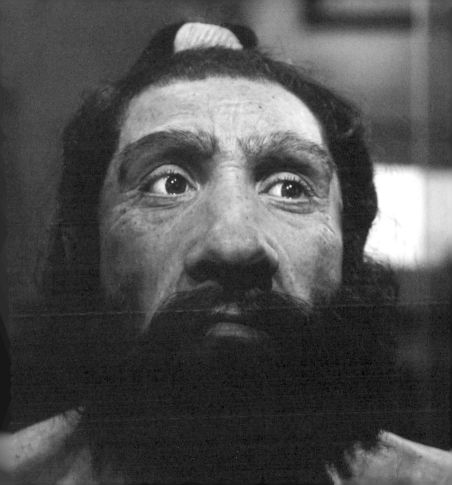

Tools and artwork

Tools, technology and artwork are staple features of human life today. Humans inherited the manufacture and use of tools from earlier hominins, many of whom had been using stone, bone and wood tools for millions of years. Most or all extinct *Homo* species, including Neanderthals, made and used tools similar to those of *Homo sapiens*. Neanderthals might also have created basic pieces of artwork. What makes *Homo sapiens* unusual is the sheer volume and detailed nature of tools and artwork produced. By about 50,000 years ago, humans were making bone needles and other delicate structures to use in the manufacture of other tools and clothing. Ancient artwork dates back to the Upper Paleolithic, between 40,000 and 20,000 years ago. It is primarily associated with western Europe, but the 2014 discovery of ancient artwork in Sulawesi indicates that the creation of rock art was widespread across the ancient world. Artwork produced by prehistoric people includes etched shells, bones and teeth, statues of humans, mammoths and horses, and spectacular cave scenes like those of Lascaux in France.

Throughout history, people have made images that convey messages, tell stories or even provide instruction on how to hunt.

The spread
of *Homo sapiens*

Our species – *Homo sapiens* – is geologically young, seemingly originating about 200,000 years ago. Fossil and genetic evidence shows that we are an African species that migrated out of the continent around 100,000 years ago, the descendants of that band of migrants giving rise to the majority of humans elsewhere. What seems most likely is that humans moved across the Arabian Peninsula and eastwards across southern Asia, later moving north and south from this general area. There are also indications that one group of our species left northern Africa much earlier, that particular group later becoming extinct.

The phrase 'Out of Africa' is somewhat misleading because it implies that the modernizing of our species was dependent on an exodus from Africa. In fact, *Homo sapiens* was fully modern long before it migrated out of Africa, plus many populations of our species never left Africa at all. Also worth noting is that some people living in Africa today descend from groups that moved there after spending time on other continents.

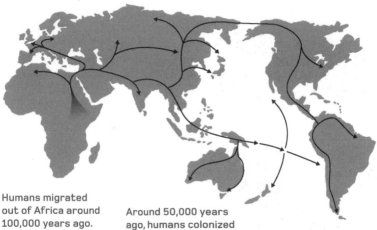

By 70,000 years ago, humans had moved east right across Asia and also invaded Europe.

By crossing the Bering Landbridge about 15,000 years ago, humans invaded the Americas.

Humans migrated out of Africa around 100,000 years ago.

Around 50,000 years ago, humans colonized Australasia from Asia.

Humans used boats to move across the vast Pacific, beginning around 3,000 years ago

The variable human

Humans are one of the greatest success stories in evolutionary history. We exist at populations exceeding those of other large animals, have colonized the planet from pole to pole, and exploit an incredible diversity of resources.

Humans are one of the most variable mammal species. We vary in skin colour, body and facial hair, height, and in the proportional length and robustness of our limbs. Clearly, we have adapted quickly to an extraordinary range of habitats: there are humans of temperate plains and woodlands, tropical forests and islands, and deserts and polar regions. Unlike other mammals, we never rely solely on our bodies, but use technology to make shelters, manufacture and shape clothing, collect food and obtain water.

Something that might help explain our variation – and thus our success – is the fact that we interbred with the other hominins that lived alongside us in the past, and have inherited many of their genes.

Urban evolution

Town, cities and other human-constructed habitats are not inhabited by humans alone, but by numerous plants, animals and other living things that have learned to live alongside humans. Many of these organisms are able to inhabit urban environments because they are highly adaptable and flexible in behavioural terms. Others, however, have adapted to these places by changing behaviourally, anatomically or ecologically. New forms or strains of these organisms have evolved into new, urban species.

The most familiar case concerns the dark form of the normally pale peppered moth, better camouflaged in polluted urban environments and thus advantaged relative to its ancestors. Adaptations to urban living have also been seen in mice, songbirds, spiders and other animals. The significance that urban places now have for many living things means that we should construct them with other organisms in mind. We need green spaces to be retained, and the keeping of ponds, trees and verges to be encouraged.

The peppered moth remains one of the most famous examples of adaptation to the urban environment. Despite claims otherwise, its validity as an example of evolution has not been disproved.

Humans:
a force of change

Humans are the most significant evolutionary force on Earth, our impact on environments, climate and even the chemistry of the land and water being greater than that of any other phenomenon. Farmed animals – different from their wild ancestors because they have been selectively bred to serve as food items – have escaped from captivity and interbred with wild populations, modifying their biology. Plants and animals moved around the world by humans also have major evolutionary impacts on native species. The influence we have on climate is affecting growth rates and distribution in animals and plants.

Increased concern over human-made climate change has resulted in a boost of technological advances and efforts to reduce carbon emission. On the other hand, a movement involving billionaire industrialists, wealthy politicians and fringe scientists seeks to deny that humans are causing global changes and aims to maintain, or even increase, our carbon output and continue the pace of ecosystem destruction and degradation.

Genetic resistance

A major area of research today concerns the resistance that bacteria have evolved to antibiotics. Antibiotics are crucial to global healthcare and involve both synthetic products and chemicals originally created by fungi and other organisms. Overuse of antibiotics and rapid evolution in bacteria have led to the emergence of resistant strains. Consequently, resistance to antibiotics such as penicillin is now widespread. People with reduced immune systems are now frequently in danger of attacks from newly evolved, resistant bacterial strains.

Similar problems affect agriculture, where mites and insects have evolved resistance to pesticides. Mosquitoes that spread the parasite *Plasmodium* – the cause of malaria – have also evolved resistance to chemicals that once used to kill them. The rapid evolution of resistance means that scientists must develop new chemicals to control those organisms that affect the health of people, and of domestic plants and animals in a never-ending cycle.

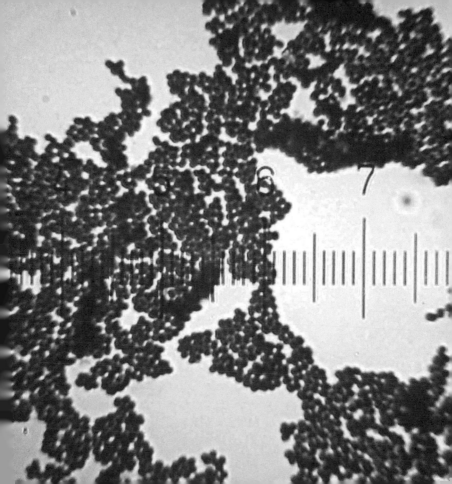

Species anew

The suggestion that species are constantly evolving requires that new ones appear regularly. Given the speed at which evolution can occur on occasion, it is not surprising that new species have emerged in the relatively short time that we have been studying them scientifically. What is especially interesting is that some of these species are unique denizens of urban and suburban environments and have, therefore, evolved as a consequence of human environmental change.

Among the most famous is a mosquito endemic to underground railways and other tunnel systems, and associated in particular with the London Underground. It is actually far more widespread and is thought to have originated in underground environments in Egypt within the last few centuries. Genetic work published in 2004 indicates that it is likely distinct enough from its ancestral population – the globally widespread mosquito *Culex pipiens* – to be regarded as a distinct species. It is also behaviourally distinct from it and the two do not readily interbreed.

Insects have fast generational
turnovers and quickly respond
to change. It is unsurprising
that some recently evolved
species, such as the underground
mosquito, belong to this group.

Species in the laboratory

Humans have modified organisms in laboratories for decades, initially by hybridizing them, more recently by modifying their genetics and by inserting synthetic components. Examples of 'laboratory-made' organisms include fruit flies that possess extra limb or body segments, and mice with modified immune systems or physiologies. At least some of these organisms are distinct enough from their ancestors to be regarded as new species. They include new forms of yeast and a hybrid lizard bred into existence in 2011.

Some organisms have been bred in the laboratory for so long that they have gradually evolved into something new. The most famous case concerns a strain of human cancer cells, collected from a young woman named Henrietta Lacks in 1951 and cultured ever since. These cells – termed 'HeLa' cells – are now very different from typical human blood cells and behave like single-celled organisms. Some experts argue that HeLa cells are now a distinct organism that should be recognized as a new species.

So-called 'HeLa' cells exist uniquely in the laboratory. Of human origin, they have been named as a new species, *Helacyton gartleri*, although this proposal is controversial.

Controlling evolution

Humans control evolution in numerous species. We have exerted massive control over domesticated plants and animals, and – via selective breeding – have created many new organisms notably different from their wild ancestors. Through hunting and culling, we control genetic variation in large mammals, birds and fishes. At least some of these efforts are (or have been) managed, the aim being to keep populations below a certain size or, in cases, remove problem individuals. Arguments can be made, therefore, that evolution in such animals as elephants, lions, deer and tuna has been controlled through human action.

The broader impact of this control is chaotic and unplanned. In some cases, the elimination of top predators has allowed smaller, lower-level predators – examples include the coyote – to become bolder, more widespread and more important. It has also resulted in the booming of herbivore populations, leading to the overgrazing and degrading of certain habitats.

By selectively removing big, showy individuals from a species, humans have affected the genetic diversity and evolution of that species, a pattern noted in the American bison.

Predicting the future

Specific future events in evolution – those involving particular acts of selection – are impossible to predict with certainty. A drought or flooding event, for example, could occur and change the selection pressures acting on a given population. However, it is also true that evolution occurs in a predictable fashion and that changes set to occur in some organisms *can* be predicted with a high degree of accuracy, at least among some groups of organisms. Studies of anoles – a group of Caribbean climbing lizards – have shown how evolution occurs predictably, again and again, when different species adapt to the same environments.

One area involving prediction in evolution concerns changing populations of cancer cells, and changes in microorganisms such as yeast. As populations evolve, they mostly follow similar, predictable evolutionary routes as they increase in fitness. This has significant implications for the health industry. It means that evolutionary pathways likely to be taken by evolving microbes and tumour populations can be pre-empted and curtailed.

Anoles are a large group of American lizards. Species within the group have repeatedly followed the same specific pattern of evolution.

Conscious evolution

Humans are the first organisms in history to be aware of the process of evolution, and to have the ability to choose which traits – anatomical, genetic and behavioural – we allow to persist into the future. Our rapidly improving knowledge of genetics is enabling us to identify which genes are problematic, and genetic testing is now common and relatively cheap for people of all ages, including the unborn. The result is that we are gradually weeding out genetic diseases and conditions. Humans in countries with healthcare and fair systems of equality also have total control over whether they reproduce at all. This means that some sectors of the human population (those without this control) contribute more to future evolution than others. The emerging picture, then, is that we are consciously shaping our own evolution. The great caveat here, however, is that such things as genetic screening and reproductive rights are unavailable to large portions of the human population. So while 'conscious evolution' might be occurring, it is not necessarily taking place across the whole of our widespread and abundant species.

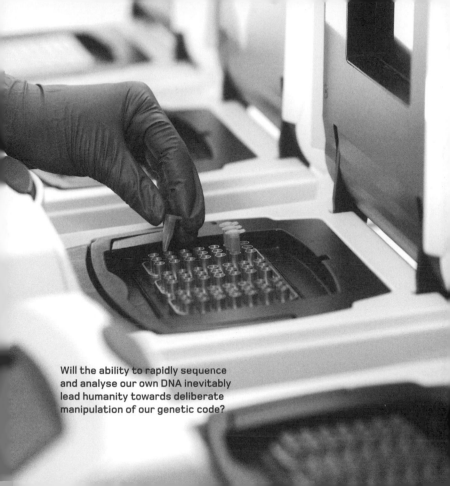

Will the ability to rapidly sequence and analyse our own DNA inevitably lead humanity towards deliberate manipulation of our genetic code?

Will humanity become extinct?

No species lasts forever. Our species has been in existence for at least 200,000 years and is not old in evolutionary terms. Consequently we should persist for hundreds of thousands of years yet; there are no indications that *Homo sapiens* is in decline.

However, whether we have the capacity for total destruction – through nuclear annihilation or such – remains debated. Some experts argue that humans are so numerous, widespread and suited to survive catastrophe that no event would result in extinction. A collapse of global ecosystems resulting from climate change and overpopulation has also been predicted as a cause for our extinction but, again, it is difficult to accept that such events could cause human die-off everywhere. Ultimately, it is impossible to predict what might happen in coming centuries. But, more so than any other species, humanity is able to control its evolution via the use of technology, genetic modification and the control of resources. These factors make it likely that we will persist for longer than has been typical for mammal species before us.

Robotics and cybernetics

The idea that humans might merge with technology is often regarded as an inevitable part of progress. Already, significant advances have been made regarding prosthetic structures that improve health and increase longevity. The number of devices that interact with our biological systems – and can thus be considered cybernetic implants – is clearly on the increase.

It is conceivable that sufficient advances in this field might allow people to prolong their lives, or to remain healthier or more mobile than might otherwise be the case. There is, however, some way to go before biological evolution and technological implants can be imagined occurring together, or when robots or implants are able to self-repair, reproduce and act like organisms. It should also be noted that such advances are mostly limited to wealthy countries, or to that minority of people who have access to a technological infrastructure. The consequence is that any such advances might not be impactful in relation to human evolution as a whole.

Artificial organisms

Humans have modified organisms through domestication, selection via hunting, and the creation of new environments. Some of these living things might be considered artificial in a sense, but they are still formed of the essential parts and components provided by their original biology. They are not organisms that have been 'created in the lab', to paraphrase an analogy from the world of science fiction.

Several efforts to create new organisms or modify the parts of existing ones are currently underway. All stem from our improving knowledge of the genetic code and how to manipulate it. Since 2014, a team led by Floyd Romesberg at the Scripps Research Institute in California has generated microorganisms that possess artificial genetic components that do not occur in nature, resulting in microbes with an artificially expanded or augmented genetic code. It is hoped that such bioengineering will eventually result in genetic combinations that allow the construction of entirely new biological functions and processes.

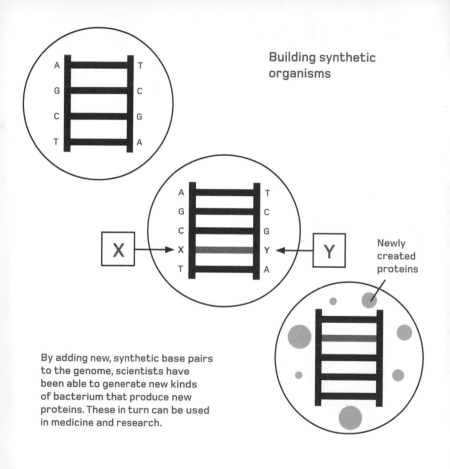

Building synthetic organisms

By adding new, synthetic base pairs to the genome, scientists have been able to generate new kinds of bacterium that produce new proteins. These in turn can be used in medicine and research.

Newly created proteins

The future of evolution

Despite all the damage we may be doing to the organisms sharing the planet with us today, it is likely that many will persist and continue to evolve. However, we cannot know which selection process will be at play in the future, nor which organisms will be around to respond to them.

Several efforts to imagine future life have appeared. In 1981, artist and author Dougal Dixon populated the Earth of the future with speculative birds, rodents, primates and other creatures for his book *After Man*. Several other authors have made similar attempts. As for what might really evolve, in the short term we might imagine a world in which diversity is low and new dynasties have arisen from widespread pest species able to live alongside humans, including rats and feral cats on the land, and fast-growing squids and jellies in the sea. Whatever form future life will take is, of course, unknown. But if there's one thing we can state with confidence, it's that living things will continue to evolve for many millions of years to come.

We will never really know which organisms will evolve in the future, but informed speculation can be fascinating. This animal – dubbed the Reedstilt – is one of numerous imaginary future animals created by author and artist Dougal Dixon.

Glossary

Anatomy
Anything pertaining to the structure in living things, or to the study of that structure.

Ancestor
Any organism (either an individual or a species) that gives rise to additional organisms, its descendants.

Carnivorous
A lifestyle where food consists wholly of animal material, and specifically the tissues of land-dwelling vertebrates.

Consciousness
The condition whereby an organism can be demonstrated to be aware of its surroundings and its own actions.

Domestication
The process in which organisms are modified to become better suited for human use.

Environment
The area around an object of interest, but most usually associated with the natural space and ecosystem of a particular region.

Generation
A set of individuals that are born or hatched at – approximately – the same time, or the period corresponding to the time during which that set of individuals mature and produce the next generation themselves.

Habitat
The region associated with a particular species and favourable for its survival and way of life.

Herbivorous
A lifestyle or organism where food consists wholly or mostly of plant material; all full-time herbivores occasionally eat animal material.

Heritable
In biology, the phenomenon in which information or characteristics can be passed down the generations.

Invertebrates
A vast group of organisms including all those species lacking bones: everything from single-celled creatures to molluscs, arthropods and echinoderms.

Microorganisms
Any of the millions of organisms that are invisible to the naked eye.

Omnivorous
Any organism that is not specialized as goes diet, but can consume both animal and plant material.

Parasite
Any organism that survives by exploiting the resources provided by another and does not provide any benefit to its host.

Physiology
The biological processes that keep an organism alive, and their study.

Plasticity
The ability of a species to modify its biology according to a number of different circumstances.

Population
Any group of living things that have a genetic connection; those members of a species inhabiting the same area, or a group of species that are being discussed together.

Proteins
The set of large, complex molecules, consisting of amino acids, that are essential to the way cells – and hence organisms as a whole – function.

Symbiotic
A lifestyle or relationship where two organisms (individuals or species) live in proximity and have a relationship where both benefit in some way.

Vertebrates
The group of organisms that possess bones: the fishes, amphibians, reptiles and mammals and their extinct relatives.

Viruses
A microscopic, abundant group of entities that are dedicated parasites of organisms; it remains debatable whether viruses themselves should be regarded as organisms or not.

Geological time periods

Time (millions
of years before
present)

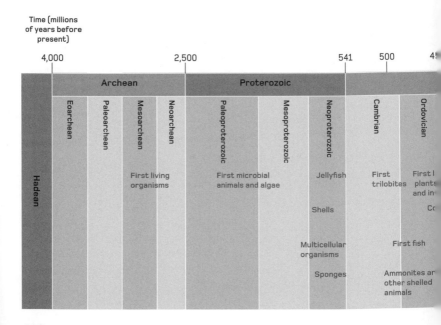

	4,000		2,500		541	500	4!

This timeline shows the major subdivisions of geological time used when discussing the history of life on Earth. Major eras, defined by significant changes in the nature of life, are subdivided into geological periods, with notes indicating major innovations and themes in the evolutionary story. Note the diagram is not to chronological scale.

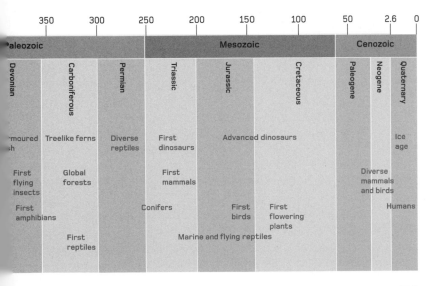

Index

Quercus

New York • London

Text © 2017 by Darren Naish
First published in the United States by
Quercus in 2017

ISBN 978-1-68144-065-1

Library of Congress Control Number:
2017942571

Distributed in the United States and Canada
by Hachette Book Group
1290 Avenue of the Americas
New York, NY 10104

Manufactured in China

10 9 8 7 6 5 4 3 2 1

www.quercus.com

Picture credits 2: Shutterstock/Bernhard Richter ; 11: Shutterstock/ Sarel; 15: Shutterstock/ Cathy Keifer; 17: Shutterstock/Victor Lapaev; 19: Shutterstock/Tsekhmister; 23: Shutterstock/Pichugin; 33: Richard Ellis/Science Photo Library; 39: Ghedoghedo via Wikimedia; 63: Shutterstock/Michael Zysman/Anton Cherkasov/Ian Kennedy/Marcel Klimko; 71: Shutterstock/Nicolas Primola; 101: Shutterstock/Di Studio; 131: Shutterstock/Marina Sterina; 133: Shutterstock/Jean-Edouard Rozey; 143: Shutterstock/Bluehand/Roman Sotola; 145: Shutterstock/ Tristan tan/Nebojsa Kontic; 147: Shutterstock/Leonid Eremeychuk; 149: Shutterstock/Andrzej Grzegorczyk; 153: Shutterstock/7th Son Studio; 155: Shutterstock/Photomatz/Pets in frames; 157: Shutterstock/Dennis van de Water/Rzeszutko/; 159: Shutterstock/Tomava; 161: Shutterstock/Hein Nouwens; 163: Shutterstock/Lenor Ko; 165: Shutterstock/Catmando/Eric Isselee; 171: Shutterstock/3drenderings; 181: Shutterstock/outdoorsman; 183: Shutterstock/Chris Humphries/Erni/ Outdoorsman/Cuson; 189: Shutterstock/Moriz; 191: Shutterstock/ Worraket; 201: Shutterstock/Iaremenko Sergii; 207: Shutterstock/ ESB Professional; 217: CDC/Dr. W.A. Clark Wikipedia; 219: WENN Ltd/ Alamy Stock Photo; 223: Shutterstock/Gio.tto; 225: Shutterstock/Mikael Häggström; 231: Didier Descouens via Wikimedia; 233: Shutterstock/ Piotreknik; 235: Shutterstock/Dennis Sabo; 237: Verisimilus Wikipedia; 239: Toby Hudson Wikipedia; 241: Jstuby Wikipedia; 251: Berengi Wikipedia; 253: Wilson44691 Wikipedia; 261: Nobu Tamura Wikipedia; 263: Shutterstock/ Albert Russ; 265: Shutterstock/Warpaint; 269: Danielle Dufault Wikipedia; 271: OpenCage Wikipedia; 279: Dr. Günter Bechly Wikipedia; 281: Gunnar Creutz Wikipedia; 283: Shutterstock/Designua; 285: H. Zell Wikipedia; 290: Kentaro Ohno via Wikimedia; 293: Dallas Krentzel via Wikimedia; 297: Kevin Walsh Wikipedia; 299: Didier Descouens Wikipedia; 301: Shutterstock/ Catmando; 303: Shutterstock/Michael D Brown; 305: Shutterstock/ Catmando; 307: H. Raab via Wikimedia; 315: Shutterstock/Seaphotoart; 315: Pterantula via Wikipedia; 317: Shutterstock/Ethan Daniels; 319: John Severns via Wikimedia; 323: PePeEfe via Wikimedia; 325: Shutterstock/ Warpaint; 327: Lisa Andres via Wikimedia; 329: Nicholas R. Longrich via Wikimedia; 331: Shutterstock/Andrej Antic; 333: Shutterstock/Johan Swanepoel; 335: Shutterstock/AuntSpray; 337: Jason Baroque; 341: Erik Terdal Wikipedia; 343: Hesupermat Wikipedia; 345: Didier Descouens via Wikimedia; 349: FunkMonk via Wikimedia; 351: Shutterstock/Dennis W Donohue; 353: Shutterstock/Mark1260423; 357: Shutterstock/Ivan Kokoulin; 361: Concavenator Wikipedia; 371: Thomas Roche via Wikimedia; 373: Rama Wikmedia; 375: John King/Alamy Stock Photo ; 377: Tim Evanson via Wikimedia; 379: Shutterstock/Zatvornik; 383: Shutterstock/Rawpixel. com; 385: Shutterstock/Kent Weakley; 388: Bob Blaylock Wikipedia; 391: Walkabout12 Wikipedia; 393: GerryShaw via Wikimedia; 397: Shutterstock/ Dennis van de Water; 399: Shutterstock/Vit Kovalcik; 401: Shutterstock/ Dotshock; 403: Shutterstock/Willyam Bradberry; 407: Dougal Dixon. All other illustrations by Tim Brown